British Museum Aircraft

39 MUSEUMS AND THE HISTORIES OF THEIR EXHIBITS

Compiled by
Ken Ellis and Phil Butler

Photos edited by Martin Chell

Published by
MERSEYSIDE AVIATION SOCIETY LIMITED
Room 14, Hangar No. 2, Liverpool Airport,
Merseyside L24 8QE

Printed by
THE ROBIN PRESS
P.O. Box 16, Brookhouse,
Peel Green, Eccles, Lancs.

MAS **A Merseyside Aviation Society Publication**

ISBN 0 902420 15 1

FRONTISPIECE

This map details the locations of the Museums mentioned in the main text. The numbers correspond to the numbers given on the Contents page, opposite.

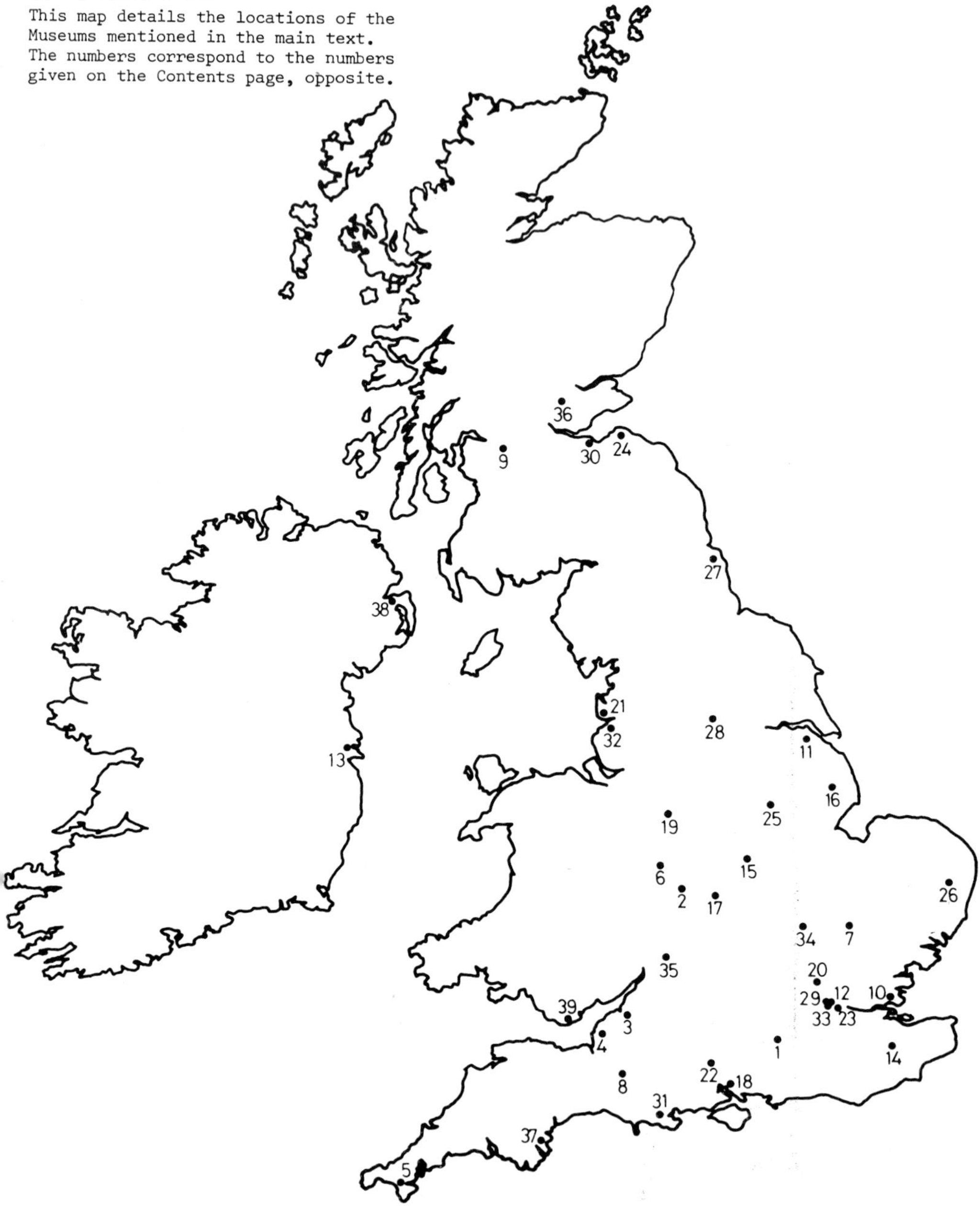

CONTENTS

The number prior to each museum below refers to its location on the map on page 2
A museum marked with an asterisk (*) failed to respond to the compilers' circulars.

Photo Pages are spread through the book at the following pages :-

16, 17, 28, 29, 40, 41, 52, 53, 64, 65, 76, 77, 88, 89, 100, 101, 112, 113, 124, 125, 136, 137, 148, 149.

PREFACE

This book stems from an idea of Phil Butler's, to fill in the inadequacies of our popular biennial 'Wrecks and Relics'. 'W & R' has always listed the contents, opening times etc. of museums in the United Kingdom and Eire, but, because of its far wider scope, has always had to restrict these to the barest minimum. Hence 'British Museum Aircraft', which we hope, treats the museums mentioned with the fullness they deserve. Here the reader can find potted histories of exhibits and the museums in which they appear.

This book is a joint effort; Phil Butler having originated the concept, undertaking specific research and acted as overall historical adviser; Ken Ellis undertaking overall compilation, research and editing and production; Martin Chell edited and laid out the photographic content; with only a few exceptions, the staff of the museums themselves have supplied the introductory narrative for each entry. Acknowledgements to the barrage of people who have helped in many ways in the production of this book appear overleaf.

'British Museum Aircraft' is very much a prototype, nothing of its like having appeared under one cover before. The compilers will therefore be extremely pleased to hear the views of readers as to its scope and to specific items in the histories presented. Any letters will be gratefully received at the address given at the end of the preface.

One of the first problems faced in producing 'British Museum Aircraft' (apart from the sheer bulk of the project) was to define a museum. We have settled on - any museum with whole aircraft exhibits, that opens its door to the public *on a regular basis*. Thus the major collections open to inspection by prior permission only are excluded; the compilers did not consider that the collections at St. Athan and Finningley, which are open once every Battle of Britain Day, were open on a reasonably regular basis, thus they are not listed in depth. Reserve collections, and the like, are mentioned at the end of each entry (where applicable) to provide a fuller picture, as they may well go on exhibition during the currency of this monograph.

The photographic content has been arranged to show exhibits *before* they entered the museum, as it is hoped the reader will actually *go* and see them as they are today.

'British Museum Aircraft' has been designed to appeal to the general reader as well as the enthusiast. To this end space saving abbreviations have been kept to a minimum so that little (if any) previous knowledge is needed to read each entry. This at times means some repetition, but we feel that it will prove its worth.

Some more specific notes on the layout of the entries follows.

All correspondence should be addressed to :-

16, Walsingham Road, Childwall, Liverpool L16 3NT.

Ken Ellis & Phil Butler 3/77

NOTES

The following notes will help the reader to extract the most from the text :-

* All entry fees, opening times etc. are published *as a guide* and the publishers cannot be held responsible for any changes there may be. The reader is advised to confirm times etc. with the museum in question before making a visit.

* Each entry contains the following :-

  1) Museum title.
  2) A brief write-up of the museum, including correspondence address etc.
  3) A listing, alphabetically by design manufacturer, of the aircraft on display in the museum at the time of writing.
  4) A section headed 'Notes' outlining incomplete airframes displayed at the museum, exhibits held in store by the museum, exhibits owned by the museum but displayed elsewhere etc.

* Each individual aircraft entry contains the following :-

  1) Aircraft designation, current identity and code (if applicable) and reference(s) to any photograph(s) of that particular aircraft inside the monograph.
  2) Date of construction, constructor's number, aircraft type, engine(s) type and power : each given only if known.
  3) History of the exhibit.

ACKNOWLEDGEMENTS

The compilers are indebted the following people for their help in the production of this monograph. The staff of the various museums are listed by their museum.

Airborne Forces Museum, Major H. M. M. Ritchie, M.C. Birmingham Museum of Science and Industry, P. Robinson - Director, and J. H. Andrew. City of Bristol Museum and Art Gallery, P. W. Elkin - Curator of Technology. British Rotorcraft Museum, Elfan ap Rees - Director, and for much help on helicopter histories, and illustrations. Cornwall Aero Park, Lt. Cdr. J. J. Leeson, R.N. (Retd.) - Exhibition Director. Duxford Aviation Society, Mike Barnett - Public Relations Officer, and Clive Norman. Fleet Air Arm Museum, Commander D. C. B. White, O.B.E., M.R.Ae.S., R.N. - Director, and Lt. Commander D. M. Rouse, M.B.E., R.N. - Deputy Director. Glasgow Museum of Transport, J. C. Clayson - Museum Assistant. Historic Aircraft Museum, W. H. Pepperell - Manager. Humberside Aviation Museum, M. R. Sargent - Curator. Imperial War Museum, Anjana Gupta - Department of Education and Publications, Martin Garnett - Assistant to the Deputy Director, J. S. Lucas - Deputy Head of the Department of Photographs. Irish Aviation Museum, Captain J. C. Kelly-Rogers - Hon. Curator. Leicestershire Museum of Technology, Peter Stoddart - Curator. Lincolnshire Aviation Museum, Graham Taylor. Midland Air Museum, John Berkeley. R. J. Mitchell Hall, Philip Morten - Curator. Mosquito Aircraft Museum, Stuart Howe, also for help on histories and many superb photos. Museum of Army Flying, J. R. Cross. Museum of Artillery, S. C. Walter - Curator. Museum of Flight, David Brown - Curator. Newark Air Museum, Peter Green - Vice Chairman. Norfolk and Suffolk Aviation Museum, R. W. J. Bullen - Secretary. North East Aircraft Museum, Patricia Traynor. Nostell Aviation Museum, Mike Cook - Chairman of SYAS. Royal Air Force Museum, Jack Bruce, I.S.O., M.A., F.R.Hist.S., A.M.R.Ae.S. - Keeper of Aircraft and Aviation Records, and Peter Murton and Richard Simpson of the Department of Archives and Aviation Records. Royal Scottish Museum, Don Storer - Department of Technology. Royal Umpire Museum, M. H. Kuiell. Science Museum, Peter Mann - Research Assistant, and Miss L. Morris of the Publications Section. Shuttleworth Collection, D. F. Ogilvy - General Manager, and H. R. Boutle. Skyfame Aircraft Museum, Peter M. Thomas - Director. Strathallan Collection, Dick Richardson, also for illustrations. Torbay Aircraft Museum, Keith Fordyce, M.A. - Curator. Wales Aircraft Museum, Gwynne Roberts - Secretary of SWAPS.

The following people have helped the compilers in specific research matters. At this point we would like to underline the help given to us by members of the Research Sections of the BRITISH AVIATION RESEARCH GROUP. BARG is highly recommended to all who do, or would like to, follow the histories of British military aircraft. This has been made even easier recently with the pioneering introduction of their specialist magazine, 'Roundel'. Membership enquiries for BARG should be sent to :-

P. Hewins, BARG, 8, Nightingale Road, Woodley, Berks. (Please enclose an SAE).

Thanks to :-

Peter Arnold for Spitfires and the Seafire histories.

British Aviation Research Group experts :-

Frank Tyler of the Army Aviation Research Section for Auster 9s and Skeeters.
And the following from the prolific Naval Research Section :-
Dave Allen for Sea Princes, W. J. Downing for modern types, E. Myall for helicopters, Doug Rough for Harvard and Avenger histories and for co-ordinating the whole effort, and to R. Sturtivant for many types.

Malcolm P. Fillmore for De Havilland Moth histories.

Paul A. Jackson, for considerable 'donkey work' on many, many histories.

Roger Lindsay for Sea Venoms, Javelins, Lightnings and some photographs.

And to the following for help in various ways in and around the book :-

Brian Billington for proof-reading; P. G. Blohm of The Robin Press; Leo and Pete Davenport; Gary Edwardson; Don Ellis for the map; Gary Finch and Tim Treherne for the cover design, with a little bit of help from Ray Jones; Rayner Kittle; Steve Langfield of Yeadon Aeroplanes; Chris Moore; Stephen Piercey of the 'Flight International' Picture Library; Brian R. Robinson of the B.A.P.C.; Mike Vines of Air Portraits; Peter Walsh; John C. Warne of Research Avion. And to everybody who contributed photographs....

AIRBORNE FORCES MUSEUM

The present Museum was set up at Browning Barracks, Aldershot, and opened by Field Marshal Montgomery in March, 1969.

It depicts the history of 'Airborne Forces' from their inception in 1940 when, inspired by a courageous and far-sighted signal from Winston Churchill, a force of five thousand airborne troops was envisaged. This signal can be seen together with the statuette of the airborne soldier and a very fine sculptured bronze of Pegasus, which is synonymous with airborne troops throughout the world.

A fine collection of hand painted figurines show the uniforms, the weapons and the equipment of the airborne forces from 1940 to the 1970s. A World War Two room shows original briefing models used by the Commanders to instruct their troops, together with a supporting detailed account of the action. Also displayed are period uniforms and weapons of American and German airborne forces. A Hotspur fuselage and a Horsa cockpit are featured in this room.

A post war room showing exhibits from the Far East and the Near East brings the Museum up to the 1970s and includes a case on Northern Ireland.

Although a young regiment by infantry standards, we possess a fine collection of medal groups showing a resume of service background of the recipient.

The Museum is open Monday to Saturday 0900 to 1230 and 1400 to 1700 hours. On Sunday from 1000 to 1230 and 1400 to 1630 hours. Admission charges are as follows :- Adults 10p, Serving and Ex-members 5p, Children 3p.

Address : The Airborne Forces Museum, Browning Barracks, Aldershot, Hampshire GU11 2DS.
Telephone : Aldershot 24431., extension 619.

Douglas Dakota C. Mk.4 KP208 'YS'

Built 1944, constructor's number 33412, , general purpose transport, powered by two Pratt and Whitney R-1830 radial piston engines of 1,200 h.p.

KP208 stands outside the Airborne Forces Museum and serves as a constant reminder of the close association between the Airborne Forces and the Royal Air Force. KP208 was built by the Douglas Aircraft Corporation at their Oklahoma City plant as a model C-47B-35-DK on order for the United States Air Force, serial number 44-77087. It was delivered to the RAF, under the designation Dakota C. Mk.4 in 1945, serial KP208. Post war it served with the Coastal Command Communications Squadron, and spent its last days of operational life as a 'hack' at Aldergrove. It last flew in 1970. It was soon ear-marked for the Museum from No.5 Maintenance Unit at Kemble and was assembled outside the barracks in 1971. It was first displayed in its Transport Command silver/blue/white scheme, but was later painted in a more representative camouflage scheme, complete with 'D' Day stripes and the coding 'YS'.

Note : As mentioned above, inside the Museum is a General Aircraft Hotspur I fuselage section and the cockpit of an Airspeed AS.58 Horsa II glider, both of which have not been identified. However it should be noted that the latter *may* be from Horsa CX. Mk.2 (post war designation) TK905 which was noted dismantled in the grounds of the Airborne Forces barracks in 1951.

BIRMINGHAM MUSEUM OF SCIENCE & INDUSTRY

The Museum was founded in 1950 to preserve examples of industrial equipment, processes and products and to develop a modern science section.

The factory premises of Elkington and Co. Ltd., a firm famous for silver work and electroplating, were bought and converted to museum galleries. A locomotive hall was erected in 1972 to house the Stanier Pacific Locomotive 'City of Birmingham' and the total display area is now more than 60,000 square feet.

Galleries include heavy engineering, transport, fire arms, machine tools, mechanical musical instruments, a science section, ornamental turning, pens and writing equipment, aircraft and replica workshops. Many exhibits operate every day while others, such as steam engines, are run on certain days each month.

Many experimental devices can be operated by visitors, including a small windtunnel in the Aircraft Section. Information on exhibits is presented in printed captions, film-strips and taped recorded commentaries. Additional information and bibliographies are available in a variety of publications sold in the Catalogue Stall.

Apart from the whole Spitfire and Hurricane the Museum also displays the cockpit section of a Bristol Beaufighter, and many engines, including a Rolls-Royce Merlin, Griffon,

Derwent, Alvis Leonides Major, Napier Sabre, Bristol Jupiter, Hercules and Pegasus, Daimler-Benz DB.605, Gnome Rhone and a BMW 109-003E axial flow engine.

Admission is free. Opening hours are Monday to Friday 1000 to 1700 hours; Saturday 1000 to 1730 hours; Sunday 1400 to 1730 hours; and the first Wednesday in each month 1000 to 2100 hours.

The Museum is five minutes walk from the Town Hall and close to the Post Office Tower.
Address : Birmingham Museum of Science and Industry, Newhall Street, Birmingham B3 1RZ.
Telephone : 021-236-1022.

Hawker Hurricane IV KX829 'JV:I'

Built 1943, single seat fighter, one Rolls-Royce Merlin XX piston engine of 1,260 h.p.

KX829 was first issued to the RAF at No.22 Maintainance Unit at Silloth on August 23rd 1943. Its first active service came with No.137 Squadron from January 21st 1944 when it was delivered to that unit's base at Milfield, later moving on to Rochford. In June it left to join No.286 Squadron, followed by No.1606 Flight in November 1944. In 1945 it moved into north western Europe for service with No.631 Squadron as part of the occupying forces, from August 1st. In 1946 it was retired from active service and came back to its original home at Silloth on March 7th for storage and disposal. Interestingly enough its last flight is recorded as having been on June 30th 1945, at which time it has amassed only 320 hours in the air. KX829 was bought by Loughborough College (later Loughborough College of Advanced Technology) and it joined an impressive array of instructional airframes at the campus. As a Mark IV Hurricane KX829 had metal frame wings. These were donated to the Shuttleworth Trust's Hurricane in order to render that aircraft more robust to external display. KX829 now wears the wooden frame wings of that aircraft. In 1961 the machine was presented to the Museum and it was painted up in the colours of No.6 Squadron, whose code, 'JV:I', it presently wears.

Supermarine Spitfire LF.IX ML427 'IS:T'

Built 1944, single seat fighter, one Rolls-Royce Merlin 66 piston engine of 1,580 h.p.

Built locally at the famous Castle Bromwich plant of Vickers-Armstrong ML427 was issued to No.9 Maintenance Unit at Cosford on April 29th 1944 for preparation for service. This came the following month, on the 16th with No.3501 Support Unit, until joining the Fighter Leader School on the last day of May 1944. In June, the 7th, ML427 again rang the changes, going to Rolls-Royce for tests at their airfield at Hucknall. On July 23rd the Spitfire was on charge at RAF Milfield. Its life from now on was to remain one of storage and test work. No.29 M.U. at High Ercall received the aircraft on August 27th 1945 and held it until October 31st the following year when it was issued back to its builders and was based at South Marston. This did not last, for September 10th 1947 saw ML427 on charge with No.6 M.U., Brize Norton, where it was officially demoted to instructional airframe status on October 15th and ear-marked for issue to No.4 School of Technical Training at St. Athan. That day ML427 relinquished its serial for the maintenance number 6457M. In August 1954 it was back to its birth place for duty guarding the gate of RAF Castle Bromwich. It was taken on by the Museum in 1958, and is displayed wearing the codes 'IS:T'.

Note : Also on display is the cockpit section of an unidentified Bristol Beaufighter.

CITY OF BRISTOL MUSEUM & ART GALLERY

The City of Bristol Museum and Art Gallery spans an extensive spread of studies, including Applied Art, Fine Art, Archaeology and History, Geology, Natural History, Oriental Art, Technology, Agricultural and Social History and is further enhanced by many other branches, including the 16th century house Red Lodge, the St. Nicholas Church Museum, Blaise Castle House Museum, Kings Weston Roman Villa, Sea Mills Roman Site, Ashton Windmill and a large reserve collection. Details of all these are available from the Museum.

In the Technology Section can be found a number of horse-drawn vehicles, high-class conveyances like the Duke of Sutherland's full-dress chariot of 1840 stand alongside more functional vehicles such as a delivery van, a gypsy wagon and a road coach 'Magnet' of 1870. Here also are the remarkable 7 mm fine scale models of the G.W.R. locomotives and rolling stock of the period 1847 to 1945. A series of ship models, mostly 1:64 scale, is being built up and those on display include Brunel's S.S. 'Great Western' and 'Great Britain'. Bristol's aircraft industry is represented by a replica of a Boxkite of 1910, which is suspended in the front hall.

Admission is free to the main Museum. The Museum is open from Monday to Saturday

1000 to 1730 hours. Closed Sundays.

Address : City of Bristol Museum and Art Gallery, Queens Road, Bristol BS8 1RL.

Telephone : 0272-299771.

Miles Boxkite replica B.A.P.C. 40

Built 1965, constructor's number BM.7281, flyable external replica for film use, one Rolls-Royce/Continental A-65 'flat-four' piston engine of 65 h.p. driving a pusher propellor.

This craft was built for the film 'Those Magnificent Men in their Flying Machines' in 1965 at Ford by F. G. Miles Ltd. It is an accurate external replica of the original 1910 design, but was built using modern construction methods. It was the third machine to be built, and was known affectionately by the builders as 'Box Three'. Although built to full airworthy standard, it is believed that this example did not fly and was only used for studio close-up shots and crash scenes. After filming it was refurbished by the builders, to the order of the British Aircraft Corporation, Filton Division, and presented to the Museum. It has since been allocated the British Aircraft Preservation Council identity number 40.

Note : The Museum has in store Bristol B.171 Sycamore HR. Mk.14 XL829, this may only be viewed by prior appointment with the Curator.

BRITISH ROTORCRAFT MUSEUM

The origins of the British Rotorcraft Museum go back to 1963 when the Director, Mr. Elfan ap Rees, then an employee of Westland Helicopters, began to assemble a unique collection of helicopter and autogyro history, including photographs, documentation and actual examples.

Since that beginning an impressive amount of material and data has been compiled and now includes histories of every British rotorcraft type and in fact of every individual machine built to date, as well as extracts from test-pilot logs and flight records, many rare photographs, and seven complete aircraft. The Museum also holds various helicopter parts including the original rotor-blades from the 1932 Hafner R.2 Revoplane. In 1973 Mr. ap Rees was appointed Consultant Historian to Westland Helicopters Ltd., and the Museum officially came into being in January 1976, as a member of the British Aircraft Preservation Council.

The rotorcraft on charge to the Museum currently comprise an early Bristol Sycamore Mark 3 (G-ALSX - at Staverton), the sole Thruxton Gadfly (G-AVKE - at Rochford), a SARO Skeeter Mark 12 (XM556), a Westland Dragonfly (WG719), a Bristol Belvedere (XG452), the sole Campbell Cougar (G-BAPS) and a Westland Whirlwind Mark 7 (XG596).

The British Rotorcraft Museum has, as yet, no display buildings of its own and the aircraft have therefore mostly been placed on loan with other Museums under the British Aircraft Preservation Council flag. The Belvedere however has returned to the Westland factory, where a group of enthusiasts are refurbishing it to static display standard. This is a particularly difficult task as many parts were missing. Several other exhibits are also now at Weston-super-Mare.

The Museum is currently negotiating for a permanent site at Weston-super-Mare which, if acquired in time,will allow a regular weekend and peak holiday opening to the public during 1977. Should this not be possible there are plans for displaying the aircraft at four Open Days during the summer and details for these will be publicised in the Aviation and local press at the appropriate time.

For further information, please write to the British Rotorcraft Museum, Delta House, Summer Lane, Worle, Weston-super-Mare, Avon, BS22 OBE, enclosing a stamped addressed envelope.

Bristol 192 Belvedere HC. Mk.1 XG452

see photo 2

Built 1960, constructor's number 13347, heavy transport twin-rotor helicopter, powered by two Napier Gazelle 100 turboshafts of 1,650 h.p. each.

XG452 was the fifth prototype Belvedere, first flying from Oldmixon on February 26th 1960. It was retained by the makers as a trials aircraft and was extensively modified to full production configuration by March of that year. It was delivered to the Aeroplane and Armament Experimental Establishment at Boscombe Down on April 27th for evaluation and trials. It flew to Gatwick on June 12th in preparation for a record flight attempt. It took off from Gatwick at 0434 hours on June 14th en route Rome in eight hours thirteen minutes, Malta in twelve hours six minutes and Idris, in Libya, in thirteen hours fifty-seven minutes. It completed tropical trials in Libya on August 16th and then flew to

Chambrey, in the French Alps, for high altitude tests. It returned to its home country on August 26th. In September it was displayed at the Society of British Aircraft Constructor's display at Farnborough. It was back at Boscombe Down for various trials from October until July 1961. The following month it returned to Oldmixon for company trials, which included paratrooping in October 1962 and radio tests. XG452 then moved to Rolls-Royce's test airfield at Hucknall in January 1963 for engine development work, including trials back at Boscombe Down again, and again at Idris. In February 1965, it came back to Oldmixon, this time for trials on its flight control system and winch. It was then refurbished and painted up in camouflage in 1967 and flew again for the RAF on August 2nd. In October it was grounded after vibration problems and allocated the maintenance airframe serial 7997M. It was delivered to No.2 School of Technical Training at Cosford on December 8th. Here it was used for radio installation instruction until 1971. The following year it was declared surplus to No.2 SoTT's requirements and disposed of to nearby Ternhill and issued to the fire section for destruction. In November 1974 it was rescued by what was to become the Museum and delivered to Oldmixon by road on November 5th 1974, where its restoration has since gone on apace under the specially formed Westland Belvedere Preservation Group. XG452 has some 487 flying hours to its credit.

Campbell Cougar G-BAPS

Built 1973, constructor's number CA.6000, single seat prototype development autogyro, powered by one Rolls-Royce/Continental O-240-A 'flat-four' piston engine of 130 h.p.

This aircraft was built as a private venture (with some backing originating in Australia) by Campbell Aircraft at Weston-super-Mare as a general purpose autogyro, with particular application in the world of agricultural aviation. It first flew on April 25th 1973 and was displayed at the Paris Salon of that year and later at the Cranfield Light Aviation Show. It flew intermittently in 1974 and 1975, before the whole project was abandoned due to lack of investment. G-BAPS has a total of 150 flying hours. It was taken on by the Museum in May 1976.

Saunders-Roe Skeeter AOP. Mk.12 XM556

see photo 4

Built 1959, constructor's number S2/5110, two seat observation helicopter, powered by one De Havilland Gipsy Major 215 piston engine of 215 h.p.

XM556 was the 38th production Skeeter, first flying at Eastleigh in July 1959. It served with No.654 Squadron, Army Air Corps, at Middle Wallop from September 1959 until June 1961. It then joined the RAF at Ternhill the following month, with the Central Flying School. It served as such until 1964 when it was reduced to an instructional airframe and moved to the Army Apprentice College at Chepstow and then on to the Royal Electrical and Mechanical Engineers School at Arborfield in December 1964 with the maintainance serial 7870M. In November 1972 it was bought by Flint Technical College and was given the boom of XM529/7979M as it was refurbished as a ground running airframe. It was bought by the Museum in April 1976 and placed on public display at the Museum's offices in Weston-super-Mare in July.

Westland WS-51 Dragonfly HR. Mk.5 WG719

see photo 5

Built 1952, line number WA/H/50, general purpose helicopter, powered by one Alvis Leonides 521/1 radial piston engine of 500 h.p.

The 50th production Dragonfly, WG719 first flew from Yeovil on September 19th 1952 as a HR. Mk.3. It was delivered to No.705 Squadron, Fleet Air Arm, at Gosport on October 7th. It stayed there until 1957 when it was converted by Westlands to HR. Mk.5 standard. It then served with the Station Flight at Yeovilton from April 1958 until that September. After a spell of modifications at Fleetlands WG719 joined No.776 Squadron, the Fleet Requirements Unit, in August 1959, until April 1961. It was reconditioned at Fleetlands in June 1961 and then served with the Britannia Royal Navy College, Dartmouth, from September 1963 to April 1967. In June it moved for storage at Fleetlands and was sold to Twyford Moors Ltd., in May 1970, moving to their base at Southampton Heliport. It remained there, in the open, until 1973. It was purchased by the Museum in November 1973 and loaned to the Fleet Air Arm Museum, and displayed at Yeovilton from April 1974 until March 1976. WG719 then moved to Weston-super-Mare for refurbishing in April and appeared briefly at the 1976 Avon Air Show. It has a total of 1,421 flying hours.

Westland WS-55 Whirlwind HAS. Mk.7 XG596

Built 1957, line number WA/H/91, general purpose helicopter, powered by one Alvis Leonides Major 755 radial piston engine of 750 h.p.

This aircraft was the eighth production Whirlwind, first flight taking place on May 6th 1957.

It was delivered to the Aircraft Handling Unit at Lee-on-Solent on the 14th and officially taken on charge with the Fleet Air Arm two days later. On that date it was issued to No.700H Squadron at Lee and coded '508/H'. It came back to the AHU in October and went back to the makers at Yeovil in early 1959 for clutch development trials. On December 1st 1960 it moved to the AHU at Fleetlands and from there to No.737 Squadron at Portland on February 9th 1961, gaining the codes '775/PO'. July the following year saw it on loan to No.705 Squadron at Culdrose, before returning to Fleetlands on September 20th. June 21st 1963 it was back at Portland, this time with No.771 Sqn, coded '519/PO'. It was back at Fleetlands on December 11th 1965 and issued on a permanent basis this time to No.705 Squadron on November 25th 1966, with the code '563/CU'. In July 1970 it suffered some damage but was back in use with the unit, this time as '556/CU' on October 17th 1971. On December 17th 1974 XG596 was retired and flown to the Royal Naval Aircraft Yard at Wroughton and used for apprentice training, receiving the maintenance airframe number A2651 and its engine the number E4699. The Museum took charge of it in December 1976 and it is intended to keep it in flying condition. Total flying time on acquisition was 3,051 hours.

Note : The following aircraft are owned by the Museum but are displayed elsewhere :-

Bristol 171 Sycamore Mk.3 G-ALSX - at the Skyfame Aircraft Museum, which see.
Thruxton HDW-1 Gadfly G-AVKE - at the Historic Aircraft Museum, which see.

CORNWALL AERO PARK

Cornwall Aero Park was purpose-built to provide all-weather family entertainment on a seasonal basis. Outside there is an impressive range of aircraft including a helicopter and a hovercraft. A large well-lit exhibition hall houses a further display of aircraft together with models and a great variety of 'aeronautica'. For the 1977 season, the Park has acquired two superb collections which will fascinate all age groups and recall many memories for the not-so-young.

This is an aircraft museum with a difference because, wherever possible, it allows its visitors to actually get into exhibits and encourages them to touch things. There is a real Shackleton cockpit through which you can pass and, if you wish, sit in the pilot's seat and handle the controls; or there is the hovercraft to climb aboard.

Lifeboats, coastguards and helicopters are part of the everyday scene in Cornwall and here there are special displays depicting the work of these life-saving services - there is the actual Whirlwind helicopter which, over several years, has rescued scores of lives from the seas and cliffs around the Cornish coasts. Also there is BBC TV's 'Blue Peter' inshore rescue craft recently retired from active service in Anglesey.

A display of relics and momentos from the two world wars are another special feature and fascinate particularly the older generation.

Cornwall Aero Park is open from Easter until the end of October and is easy to find. Just head for the Helston side of Culdrose Naval Air Station which is on the main Lizard road, following the signs and park your car in the large free car park provided.

Address : The Exhibition Director, Cornwall Aero Park, Culdrose Manor, Helston, Cornwall TR13 OGA.

Telephone : Helston 3404 or 4549.

Bristol B.171 Sycamore HR. Mk.14 XJ917 'S-H' see photo 6

Built 1956, general purpose light helicopter, one Alvis Leonides 73 piston engine of 520 h.p.

XJ917 was ready for collection from Filton on December 19th 1956. By January 11th 1957 it had journied to No.20 Maintenance Unit at Aston Down for storage and checking prior to issue to No.275 Squadron on May 15th. This was to be one of only two brief sorties into service. On June 17th 1960 it was retired to No.15 MU at Wroughton, only to move back to No.20 MU by August 18th, and then to No.5 MU, Kemble, on September 19th. This period of shuttling between MU's was to finish on September 21st when XJ917 arrived back at Wroughton for preparation to enter service again. On December 16th of that year XJ917 was issued to the Central Flying School's rotary division, based at Ternhill, where the code 'S-H' was placed on the tail-boom, and remains there to this day. XJ917 was retired on June 9th 1966 and went to Wroughton for a long period of store. On September 6th 1971 XJ917 was declared surplus. (On April 5th 1972 No.15 MU officially became part of the Fleet Air Arm's domain, and was renamed Royal Naval Aircraft Yard, Wroughton.) On December 12th 1972 XJ917 was officially struck off charge as sold to Mr. Brian C. Peters, who had bought the helicopter for the planned Aero Park. In 1976 XJ917 was put on display at Helston.

Douglas Skyraider AEW. Mk.1 WV106 '427/CU'

Built 1950, airborne early warning aircraft, powered by one Wright Double Cyclone R-3350-26WA radial piston engine of 3,300 h.p.

WV106 started off life as an AD-4W Skyraider for the United States Navy, with the serial number 124086, it flew for the first time in April or May 1950. 124086 stayed in the service of the USN until 1952 when it was prepared for shipment to the United Kingdom, this taking place in January 1953, courtesy of H.M.S. 'Perseus'. By February 20th 124086, now WV106, had arrived at Royal Naval Air Station Abbotsinch and was placed in store. On August 13th 1955 WV106 was moved to RNAS Donibristle for continued storage. In 1958 it was picked for active service and went to RNAS Culdrose on March 13th. By October WV106 was part of No.849 Squadron 'D' Flight with the coding '427/C'. The following October or November saw WV106 at the Naval Air Radio Installation Unit at RNAS Lee-on-Solent for the fitting of UHF gear. It left Lee for No.849 again, but this time with the Headquarters Flight, with the code '416/C' by February 1960. This active spell was short-lived and the 'Spad' was withdrawn from use on September 23rd 1960, the type slowly making way for the Gannet. WV106 was moved to Culdrose (next door to the site of the museum it was to join in 1975) and put into store. The aircraft should have been scrapped, but was saved by the Station's staff and refurbished to its original colours. In this superb condition, it was loaned to the Aero Park in 1975.

Hawker Sea Hawk FB. Mk.5 WM983

Built 1954, single seat fighter bomber, one Rolls-Royce Nene 103 turbojet of 5,400 lb s. t.

WM983 was, like virtually every Sea Hawk, built by Armstrong Whitworth Aircraft at Baginton, as an FB. Mk.3. Its early life was involved with development work, and for this purpose it was taken on charge with the Aeroplane and Armament Experimental Establishment at Boscombe Down on July 15th 1954. By July 1955 it had moved to the Royal Naval Aircraft Yard at Fleetlands and by January 1956 had moved on to the other end of the country at Royal Naval Air Station Lossiemouth. The move to Lossie' was in preparation for its embarkation on the aircraft carrier H.M.S. 'Centaur' for a Mediterranean and Far East cruise. For this cruise WM983 is thought to have been on charge with No.811 Squadron - the colours of which this aircraft now wears in the Air Park. In July 1956 it was back at Fleetlands where it was up-graded to a Mk.5, this work being finished by October 1st. By January 1957 WM983 was back at Lossiemouth, where it was the victim of a take off accident on March 6th 1957. By July WM983 was back at Fleetlands and issued to the Fleet Requirements Unit at Hurn on September 5th 1958. No code was allocated with the FRU. On July 27th 1959 WM983 was back at Lossie', and in the same month on to Abbotsinch. On July 27th 1961 the Sea Hawk was allocated the maintenance serial A2511 and later joined the School of Aircraft Handling at Culdrose as 'SAH-6'. It became part of the Aero Park collection in 1974.

Hawker Sea Hawk FGA. Mk.6 XE368

Built 1955, single seat ground attack fighter, one Rolls-Royce Nene 103 of 5,400 lb s. t.

XE368 was also built by Armstrong Whitworth and was taken on charge by the Fleet Air Arm on April 25th 1955. It was at Royal Naval Air Station Stretton during June and July and was seen at RNAS Brawdy in January 1956, and Royal Naval Aircraft Yard Abbotsinch in July 1956. By November it was part of No.899 Sqn, receiving the tail code 'J' and a side code in the '4..' sequence. It was noted at Lossiemouth from January 1957 until July 1958 - possibly in store. After this it may have joined No.803 Squadron and received the coding '145/J'. By January 1959 it was at RNAY Fleetlands and was to remain there until January 1960. It appeared at Abbotsinch again in July of that year, leaving in January 1961. By July it had joined No.738 Squadron as '738/LM' and stayed with the unit until July 1962. On October 10th it became an instructional airframe and was allocated A2534. It joined the School of Aircraft Handling at Culdrose and was coded with them as 'SAH-3'. At some point in its career with the SAH it was given the outer wings of fellow WM913. Refurbished, in the colours of No.897 Squadron with Suez ('Operation Musketeer') campaign strips, it joined the Aero Park in 1975.

Westland WS-55 Whirlwind HAR. Mk.1 XA870

Built 1954, general purpose helicopter, one Alvis Leonides Major 755 of 750 h.p.

XA870 first flew from Westland's Yeovil plant on May 18th 1954 and was delivered to the Royal Naval Aircraft Yard at Gosport on July 19th. On September 9th it was sent by sea to the Far East and was received at Sembawang on October 21st. Two days later it was taken on strength with No.848 Squadron. It was lent to the RAF, No.155 Sqnuadron, from March 17th until June 8th 1955. It was despatched home by sea on July 4th and was received back at Gosport on August 9th. On October 1st it became part of H.M.S. 'Protector's' flight for duties in the Antarctic. With 'Protector' it was coded '910' (other sources quote '991'). XA870 stayed with the flight, apart from a brief spell at Lee-on-Solent from June to September 1956, until moving to Lee on May 22nd 1957. Here it was prepared for service with No.705 Squadron on loan from August 26th 1957. On January 20th the following year XA870 officially became part of No.705 and received the code '528' in that month. On March 11th 1959 it was heavily damaged in a flying accident, but was repaired. By January 4th 1961 it was being fitted with UHF/VHF radio installations in preparation for embarkation on 'Protector' again for another sojourn in icy climes, it officially becoming part of the Ship's Flight on June 6th. However on July 17th a forced landing brought XA870 home to a long history of being just a spare for the Flight. On July 18th 1963 minor damage was added to the bruises already taken. Finally, on May 18th 1966 XA870 was declared an instructional airframe with Lee and given the serial A2543 with the Air Engineering School. It was struck off Navy charge on May 22nd 1973 (or thereabouts) and moved to RNAS Predannack for storage prior to becoming part of the Aero Park. It is displayed in a partially sectioned form.

Westland WS-55 Whirlwind HAR. Mk.9 XN258 '589/CU'

Built mid-1950s, general purpose helicopter, powered by one Bristol Siddeley Gnome 103 turboshaft of 720 h.p.

The early history of XN258 (as a piston engined HAR. Mk.7) is very vague, it is known to have served at Royal Naval Air Station Lossiemouth as '934/LM' from at least 1963 until July 1964. On October 20th 1964 it went to Westlands at Weston-super-Mare for turbine conversion to HAR. Mk.9 standard. In this guise it first flew on December 3rd 1965, and, after testing, was delivered to RNAS Culdrose on February 7th 1966. On July 7th its joined the Search and Rescue Flight of the aircraft carrier H.M.S. 'Hermes', coded '336/H'. By September 1967 it was back at Culdrose, and in use with the Station Flight for trials on its Gnome 103 powerplant. By October 19th it was transferred to the Naval Aircraft Storage Unit at Culdrose and prepared for issue on a temporary basis to the Flight of H.M.S. 'Endurance' from July 21st 1968, until returning to NASU on September 2nd. It was to remain a local aircraft, as on July 22nd 1969 it joined the Station Flight, coded '589/CU' and was to serve faithfully as the area's aerial lifeboat until grounded by severe corrosion on January 30th 1975. On December 17th 1975 it was officially transferred to the Aero Park.

Note : Also on display is the cockpit section of Shackleton T. Mk.4 WG511 and Cushioncraft CC-7 XX101, a hovercraft in Army colours, constructor's number 001.

COSFORD AEROSPACE MUSEUM

What is now a very impressive, and full-time, Museum at Cosford started off life as a store for part of the Royal Air Force Museum's Reserve Collection. Cosford is still essentially this, but from the beginning of the 1977 season will be open to the public weekly instead of on a regular open day basis as previously. As such, Cosford is the first 'satellite' of the RAF Museum, and it is to be hoped that this trend continues.

The Aerospace Museum is named as such to take into account the very fine collection of German World War Two rockets and missiles which have been collected over the years. This collection is also open to the public.

In addition to the two main exhibitions, the aircraft and the missiles, there are several other displays to back up the hardware, these include RAF V.C.s, the work of the resident No.2 School of Technical Training, and displays by the Air Training Corps.

To cater for young children some amusements are provided and there is a restaurant available for light refreshments and in addition the Museum shop provides models, books, badges etc.

The Museum has witnessed considerable expansion in 1976, principally in taking in many of the exhibits from the axed Colerne collection, but also from other sources.

At the time of writing exact opening times and admission fees are not available, details can be had from the following address :

The Curator, Royal Air Force Cosford Aerospace Museum, Royal Air Force Cosford, Wolverhampton, West Midlands WV7 3EX.

AVRO 685 York C. Mk.1 'MW100' (G-AGNV)

Built 1944, constructor's number 1223, long range transport, four Rolls-Royce Merlin T.24 piston engines of 1,620 h.p. each.

This aircraft was initially ordered by the RAF and was allocated the serial TS798. However it was quickly diverted to British Overseas Airways Corporation, to whom it was registered as G-AGNV on December 9th 1945. With BOAC it was to serve, firstly as 'Morville' and later as 'Middlesex', on many of the Corporation's pioneering post war routes. In April 1955 it was retired and sold to Skyways Ltd, who used G-AGNV for ten years on inclusive tour and charter work. On October 9th 1965 G-AGNV, still in Skyways colours made its last flight, into Staverton, to become part of the Skyfame Aircraft Museum. With Skyfame G-AGNV acquired the first of two inappropriate serials that it has worn to date. Here it was camouflaged and painted as LV633 'Ascalon' - to represent the York C.(VIP).Mk.1 used by Winston Churchill during the war. 'LV633' stayed on display at Skyfame until 1973, and proved to be a very popular attraction during that time. In 1973 it was acquired by the RAF Museum, as one of only two Yorks still surviving in the United Kingdom. It was moved to the RAF transport base at Brize Norton and a complete restoration was initiated. Here it gained the identity 'MW100' - another VIP transport York, but still inappropriate to this particular aircraft. The restoration work was not completed however and the York was moved in late 1975 by road to Shawbury for storage, along with several ex-Colerne aircraft for eventual issue to Cosford. It arrived at Cosford in the early days of 1976 and was assembled before January was out.

AVRO 694 Lincoln B. Mk.2 RF398

see photo 8

Built 1945, long range heavy bomber, four Rolls-Royce Merlin 85s of 1,750 h.p. each.

RF398's history is one of a considerable period of storage in a considerable number of MUs, with very little actual operational service to its credit. From birth RF398 was issued to No.46 Maintenance Unit on September 13th 1945. It was with Armstrong Whitworth for a series of modifications from November 16th 1946 until January 16th 1947. It moved on to No.15 MU at Wroughton on April 25th and was then officially recorded as being on the charge of the English Electric Company from October 28th 1947 until January 5th 1948 for modifications - just what these were the compilers would like to hear. Three days later it touched down at No.38 MU at Llandow for more storage. It again visited Armstrong Whitworth from February 26th 1948 until May 5th 1948. Here the inconsistencies of the 'official' histories of aircraft can be well high-lighted - according to the records RF398 arrived at No.5 MU at Kemble two days prior to it leaving Armstrong Whitworth's! RF398 obviously took a shine to A-W as it returned for still more modifications on August 3rd 1949 until November 30th 1949. There then followed a protracted period of changing MUs, as follows : No.5 MU from December 8th 1949, No.51 MU at Lichfield from November 17th 1950, No.5 MU from August 2nd 1951, No.51 MU from November 19th 1951 and No.5 MU from May 25th 1954. Then, on November 27th 1957, came the call to service, with the Bomber Command Bombing School at Lindholme. RF398 served as such until October 4th 1960, when it arrived at No.23 MU at Aldergrove. However storage here was only transitory, as it joined the Central Signals Establishment at Watton (officially on charge with No.151 Squadron) on July 28th 1961. Its last flight came on April 30th 1963 when it was flown into Henlow and officially struck off charge. It was the centre-piece at the Royal Review in 1968 and shortly afterwards was moved to Cosford.

AVRO 698 Vulcan B. Mk.1A XA900

Built 1957, long range nuclear bomber, four Bristol Siddeley Olympus 104s of 11,000 lb s.t.

XA900 was awaiting collection by the Royal Air Force at Woodford on March 20th 1957. It was delivered three days later to No.230 Operational Conversion Unit, a unit with which it was to maintain strong ties. On June 22nd 1960 it joined No.101 Squadron for a spell of operational flying in the real sense. It was back with No.230 OCU on February 6th 1962, but only for fifteen days as it was flown to Woodford to be converted to the up-rated B. Mk.1A standard, this work being completed on August 24th of that year. On that day XA900 re-joined No.230 OCU, this time for its last flying task. On November 19th 1965 it was flown into Cosford and was reduced to ground instructional airframe status, serial 7896M, for use with the resident No.2 School of Technical Training. It joined the Aerospace Museum in December 1975.

AVRO 707C WZ744

Built 1953, two seat delta wing research aircraft, one Rolls-Royce Derwent 8 of 3,600 lb s.t.

This aircraft was one of a small family of aircraft built by Avro's to undertake research into the problems of delta-wing aircraft in the hope that they could 'iron out' problems on the Vulcan programme. As it turned out the Vulcan came to fruition quicker than planned, and the 707 series were slowed down by development problems of their own. WZ744 was a two seat side-by-side trainer version of the basic single seat 707 and first flew on July 1st 1953. After a period of development work with Avro's it then went on to aerodynamic research of a less specific nature. It was withdrawn from use in 1966 and allocated the maintenance airframe number 7932M and installed with the RAF Museum Reserve Collection at Finningley. In the early 1970s it was in store at Finningley's own 'satellite' at Topcliffe, until moving on to join the growing collection of research aircraft at Cosford.

BAC TSR-2 XR220

Built 1965, production number XO-2, two seat high speed attack aircraft, powered by two Rolls-Royce Olympus 22R-320s

The beautiful, and politically explosive, TSR-2 needs little introduction. This example was due to have flown for the first time on April 2nd 1965, but was prevented from doing so by some minor faults. Four days later, as XR220 was being cured, the entire TSR-2 project was cancelled, robbing the country of a huge potential money-earner, and robbing XR220 of its taste of flight. The aircraft was used by the Aeroplane and Armament Experimental Establishment at Boscombe Down for ground running trials of its Olympus engines - the same type that would power the equally controversial Concorde. These trials went on through 1965 and into 1966. After this it was taken to Henlow for storage, and hopeful display some time in the future when it would not be such a hot potato. On May 4th 1975 XR220 (now allocated 7933M) was taken by road to Cosford and placed on public display.

Bleriot Type XI

Assembled 1975 from original (circa 1909/10) parts, single seat monoplane, assumed fitted with an Anzani 'vee' twin cylinder piston engine of about 25 h.p.

This aircraft was part of a collection of parts bought in France by the late Leslie Goldsmith, bought originally as spares for his flyable Type XI, G-AVXV. He based the flyer with the Colerne Collection, and later donated to them on the understanding that the spares would be assembled into whole aircraft. Very little is known of the lineage of this hybrid, one other of which was built and is displayed in the Mosquito Aircraft Museum. G-AVXV is currently with the Reserve Collection at St. Athan. This example was delivered to Cosford by road from Colerne on April 27th 1976.

Bristol T-188 XF926

Built 1963, constructor's number 13519, high speed research aircraft, powered by two De Havilland Gyron Junior DGJ.10 turbojets of 10,000 lbs s. t. each.

The T-188 'Flaming Pencil' was designed in the mid-1950s for research into high speed flight, and was intended to maintain a steady Mach 2 plus. For this reason it was largely built out of stainless steel. XF926 was the second flying aircraft and flew for the first time on April 29th 1963. The T-188 project was dropped fairly quickly after the two examples had flown, and XF926 had a very brief flying career before being delivered to the Proof and Experimental Establishment at Foulness. In all it made some 51 flights, the last one being in January 1964. During one of these it managed to achieve Mach 1·88 at 36,000 feet. At Foulness is was designed to be a gunnery target, but happily it was saved from this fate and moved to Cosford for preservation.

Consolidated PBY-6A Catalina L-866

Built 1945, long range patrol amphibian, powered by two Pratt and Whitney R-1830-92 radial piston engines of 1,425 h.p. each.

L-866 first flew on April 28th 1945 as part of an order from the United States Navy, who gave it the serial number 63993. It served with the USN until December 18th 1957 when it was handed over to the Danish Air Force, firstly with the serial 82-866, later changed to L-866. Whilst with the Danes, it used the radio call-sign 'OVJBF'. It served all its life with No.721 Squadron (ESK-721) nominally based at Vaerløse, but mostly serving around the coasts of Denmark's other interest, Greenland. L-866 was damaged in collision with an ice-berg off the coast of Greenland sometime in 1968 and it was retired officially in 1969. After much deliberation as to how best to dispose of such a craft, L-866 was acquired by the Royal Air Force Museum. Although not strictly representative of the Catalinas operated by the RAF, it was deemed that one Catalina with different nose profile etc., was far better than no Catalina at all! L-866 flew into Colerne on May 29th 1974 and stayed as a major attraction with that collection until the closure of both Museum and base. It was roaded to Cosford, courtesy No.71 Maintenance Unit, on October 9th 1975.

Consolidated Liberator VI KN751

Built 1945, constructor's number 6907L, maritime patrol aircraft, powered by four Pratt and Whitney R-1830 Twin Wasp radial piston engines of 1,200 h.p. each.

KN751 started life as a B-24L-20-FO, at the Willow Run, Detroit, plant of the Ford

Motor Company. The aircraft was planned as lend-lease material for the United Kingdom and was taken to Romulus Air Force Base, Michigan , to await collection by the RAF. It was picked up, as KN751, on June 4th 1945 by a crew from No.21 Ferry Group, RAF, and flown across the Atlantic. KN751 was not through with long flights however. On June 6th it was allocated to No.99 Squadron, who were then based at Debelia, in Bengal, India. With this unit KN751 was initially used as a spare aircraft, but later saw action on detachment in the Cocos Islands and was employed on bombing missions into Burma. On April 11th 1946 KN751 was officially struck off RAF charge at Carnpore. KN751 had been sold to the Indian Air Force. The IAF required the aircraft for maritime patrol duties, a task for which the Liberator was ideally suited, and it was converted to this role by Hindustan Aeronautics, HAL completing the job in 1948. KN751, now wearing the IAF serial HE809, joined No.6 Squadron, based at Poona, and was to complete twenty years of service patrolling the Indian coastline. The Liberator was retired in December 1968 and was put into open store at Bangalore. Several Liberators from No.6 Squadron were to find buyers, principally in the United States, the Indian Government decided to present HE809 to the Royal Air Force Museum and it was readied for the long flight back to the United Kingdom. The route flown was Bombay-Abu Dhabi-Jeddah-Cairo-Rome and KN751 touched down at Lyneham on July 4th 1974, having caused a sensation by overflying the air display at Greenham Common as a superb un-programmed event. Lyneham was only a customs stop and, after some brief repairs, KN751 landed at its new home, Colerne, seven days later. With the closing of Colerne, the aircraft was dismantled and removed by road on January 20th 1976 for display at Cosford.

De Havilland DH.98 Mosquito TT. Mk.35 TA639

see photo 9

Built 1945, high speed target tug, two Rolls-Royce Merlin 114A of 1,710 horse-power each.

TA639 was built at Hatfield in early 1945 as a B. Mk.35 bomber variant, being taken on charge by the RAF in April 1945. It saw little use in this guise and was converted to target tug status, gaining the designation quoted above. In September 1954 it joined Aldergrove Target Tug Flight, serving until June 1958, when it re-entered store at No.27 Maintenance Unit, Shawbury. By October 1955 it had joined No.3 Civilian Anti-Aircraft Co-operation Unit at Exeter, and was coded '55'. On June 7th 1963 it was delivered to Little Rissington where it was taken into the care of the Central Flying School. Here it was stripped of its target towing gear and camouflaged. It was flown by the CFS under the aegis of it being a 'historic aircraft' until it was allocated the maintenance airframe number 7806M. With its flying days now ended, TA639 could boast only a total of 577 hours airborne. It has been with Cosford since its retirement.

De Havilland DH.100 Vampire F. Mk.3 VT812 'N'

Built 1947, single seat fighter, one De Havilland Goblin turbojet of 3,100 pounds s. t.

VT812 was ready and waiting for collection by the RAF on October 24th 1947 and was flown to No.5 Maintenance Unit at Kemble on the 24th of November. It was allocated for service in the Mediterranean and Middle East and flown out via Manston on August 19th 1948. It joined No.32 Squadron on December 30th, returning to the United Kingdom on July 5th 1950. It then joined the Royal Auxiliary Air Force, firstly with No.614 Squadron from July 12th 1950 and then with No.601 Squadron (in whose colours VT812 is displayed) from January 14th 1952. With No.601 VT812 suffered a Category 3 (Repair beyond unit capacity) accident on April 4th 1952. The repair was undertaken firstly by De Havilland's on site from April 21st until December 11th, and from the following day by No.58 Maintenance Unit until June 22nd 1953. It re-joined the R.Aux.A.F. with No.602 Squadron on August 8th only to fly on to No.48 MU at Hawarden on November 23rd 1953. It was allocated to ground instructional use on April 20th 1955 as 7200M and was moved to Cardington on May 18th and officially struck off charge. It joined the Colerne collection in 1964. With the closure of Colerne, VT812 was roaded to Shawbury on August 21st 1975 and moved for display at Cosford in April 1976.

De Havilland DH.100 Vampire FB. Mk.6 J-1172

Built mid-1950s, single seat fighter bomber, one De Havilland Goblin D.Gn.3 of 3,350 lbs s.t.

This aircraft was retired from long service with the Swiss Air Force in 1974 and subsequently acquired by the Royal Air Force. At the time of writing little is known of its service history. It was lastly in store at Dubendorf and was flown from there direct to Colerne on July 29th 1974. Like many other Colerne inmates, it was taken by road to Shawbury (on September 8th 1975) for storage prior to display at Cosford. It arrived at Cosford in April 1976.

1 *Sycamore Mk.3 G-ALSX in flight, now with the BRM, displayed at Skyfame* Avia Press

2 *Belvedere XG452 at Gatwick 13/6/60 ready for its record flight to Libya* Avia Press

3 *The prototype Beagle 206, now at Duxford, near Shoreham 29/8/61* 'Flight' 42060 S

4 *Skeeter XM556 'V' at Ternhill 14/6/62* P. H. Butler

5 *WG719 at the Yeovil Open Day May 1975* Avia Press

6 *Sycamore XJ917 at Ternhill 14/9/63* P. H. Butler

7 *The ugly H.126 at Le Bourget, Paris, 1965* P. H. Butler

8 *Lincoln RF398 in the static at Gaydon 9/60* P. H. Butler

9 *TA639 minus TT mods at Ternhill 18/9/65* P. H. Butler

De Havilland DH.112 Venom FB. Mk.4 WR539 'F'

Built 1955, single seater fighter bomber, powered by one De Havilland Ghost 105 turbojet of 5,150 lbs s. t.

WR539 is a recent addition to the Cosford Aerospace Museum, having been erected in late 1976. It was previously on display in the colours of No.28 Squadron inside the RAF section of Kai Tak Airport, Hong Kong. Whilst displayed there it was earmarked for the Royal Air Force Museum Reserve Collection, it was freighted home in mid-1976 and allocated for display at Cosford, still in its immaculate yellow and black No.28 Squadron colours.

Douglas Dakota C. Mk.IV 'KG374' (really KN645)

Built 1944, constructor's number 33335/16587, general purpose transport, powered by two Pratt and Whitney R-1830 radial piston engines of 1,200 h.p. each.

KN645 was ordered as a C-47B for the United States Army Air Force, serial number 44-77003. It was diverted for use by the RAF and arrived in the United Kingdom on May 27th 1945. On June 2nd it was issued to the Second Tactical Air Force Communications Squadron and started a long career in communications. On September 6th it was transferred to the British Air Forces Overseas Communications Wing and was to stay with them until 1949, apart from periods of servicing, modifications and use by No.151 Radio Unit (e.g. November 1945). On June 7th 1951 KN645 was retired for a period of store at No.12 Maintenance Unit at Kirkbride. On January 4th 1954 it was allocated to Headquarters Malta, and was on charge with the Malta Communications Flight seven days later. It returned to the U.K. on March 26th 1955 and arrived at No.22 MU, Silloth, two days later. On August 8th 1955 it joined the Supreme Headquarters Allied Powers in Europe communications flight, based at Northolt. This was but a brief spell of use, as KN645 returned to No.22 MU on December 30th and moved on to No.23 MU at Aldergrove on October 19th 1956. On January 13th 1961 KN645 was chosen to become part of the Queen's Flight, but this regal role was curtailed in March when it re-joined SHAPE at Northolt. September 1st 1962 saw it at No.60 MU Leconfield for storage, prior to issue to Air Forces North Headquarters, at Northolt, on December 18th 1962. KN645 was moved to Prestwick on May 23rd 1966 for the repair of a Category 4 (beyond repair on site) accident by Scottish Aviation Ltd. It rejoined Air Forces North in September. In September 1967 it was briefly loaned to the Ministry of Technology, and, apart from the odd visit to SAL, remained with Air Forces North until flying into No.5 MU at Kemble on April 3rd 1970. It was declared a non-effective airframe on January 10th 1972 and painted up in its present colours by No.5 MU ready for its move to the Colerne museum. It was painted as KG374 to represent the aircraft of No.271 Squadron in which Flight Lieutenant David Lord won his V.C. over Arnhem in 1945. In its new guise the Dakota was moved to Colerne by road on May 1st 1974, until the closure of that collection and its move north to Cosford on October 16th 1975 and re-assembly for display.

English Electric Canberra B(I). Mk.8 WT346

Built late 1950s, two seat bomber/interdictor, two Rolls-Royce Avon 109s of 7,500 lbs. s. t.

WT346 first saw service with No.88 Squadron and was still with the unit when it was re-numbered No.14 Squadron on December 17th 1962, by which time WT346 was coded 'F'. Then followed service with No.3 Squadron, this time as 'S' and finally with No.16 Squadron at Wildenrath in West Germany. It flew into Colerne on its last flight on June 17th 1962, with a total flying time of 4,660 hours. It was moved to Shawbury on September 24th and held in store ready for when Cosford could accept it. It finally was installed in April 1976.

Fairey Type V FD-2 WG777

see photo 41

Built 1956, constructor's number F.9422, high speed research aircraft, powered by one Rolls-Royce RA.14 Avon turbojet.

WG777 was the second 'Delta Two' to be built, and flew for the first time from Boscombe Down on February 15th 1956 in the hands of Peter Twiss. During this first flight it went trans-sonic. It was used exclusively for research flying, mostly from the Royal Aircraft Establishment's Thurleigh base. It made in all 429 flights before being withdrawn from use in 1957. It was retained as being of historical significance (although it is a pity that this machine could not have been converted into the BAC-221 test-vehicle and not the more significant WG774, now in the Museum of Flight, which see.) and kept by the Station Museum at Finningley. It was later kept at that station's 'satellite' store at Topcliffe. During its sojourn with Finningley it was allocated the maintenance number 7986M. It was moved to Cosford in the early 1970s to join the growing collection of test and research types.

Folland Fo.141 Gnat F. Mk.1 XK740

Built 1956, single seat light-weight fighter, one Bristol Siddeley Orpheus 701 turbojet of 4,700 pounds static thrust.

This aircraft was used for engine development work with Bristol Siddeley, based at Filton for all of its flying life. The single seater Gnat was not ordered into RAF service and XK740 was officially on the roll of the Ministry of Supply. It was retired and was used as a demonstration machine, based at Church Fenton. This was to continue to be its role, being taken on by No.71 Maintenance Unit at Bicester and used for several years as a mobile display machine. It was later to be repainted in the colours of the 'Red Arrows' aerobatic team, although this variant of the Gnat has never seen service with them. It was delivered to Cosford in late 1974 for use with the Aerospace Museum.

Gloster F.9/40 Meteor DG202/G

Built 1942, single seat prototype aircraft, powered by two Rover W2B-23 turbojets.

DG202/G (the 'G' suffix denotes a high security aircraft, to be guarded at all times) was the prototype Meteor, but was not the first to fly, as it was fitted with the W2B engine which proved somewhat troublesome. This honour went to DG206/G. DG202 was built in considerable secrecy, and was originally referred to as the 'Rampage'. It was ground run on June 29th 1942 and moved to Newmarket Heath where taxi trials (planned to lead to first flight) began on July 22nd. Due to engine problems DG202/G did not fly until April 1943, when it was moved on to Rolls-Royce at Hucknall for engine testing. The port engine blew in mid-air in December 1944 and this caused considerable damage. At this time it had a total of 365 hours in the air. It was rebuilt, and flew to Abbotsinch on August 11th 1945. Here it was taken to the Clyde by road and loaded onto H.M.S. 'Pretoria Castle' for deck handling trials. After this it was stored with Glosters at Moreton Valance before being taken on charge for instructional training (as 5758M) with No.5 School of Technical Training at Locking in April 1946. By January 1951 it was taken to Yatesbury and was in a bad state of dereliction by the early 1960s. It was rescued as a very significant aircraft and came to Cosford in 1965, it has since been completely refurbished.

Gloster Meteor F. Mk.8 (Modified) WK935

Built originally in the mid-1950s, two seat research aircraft, powered by two Rolls-Royce Derwent 8 turbojets of 3,500 pounds static thrust.

This aircraft was converted from a standard Meteor F. Mk.8 single seat fighter to undertake research into the possibilities of flying an aircraft from the 'prone' (lying face down) position. This was done by adding a specially built cockpit to the nose, the conversion being designed and undertaken by Armstrong Whitworth Aircraft. The aircraft had a full set of controls arranged around the pilot's couch in the new nose, but a second pilot always flew in the conventional cockpit in case of difficulties. The trials were conducted for the Bristol Aircraft Company and for the Institute of Aviation Medicine at Farnborough. After its brief test career WK935 was retired and allocated the maintenance airframe number 7869M and displayed at Colerne. It was taken by road to St. Athan October 6th 1975 on the closure of Colerne. It was moved to Cosford early in 1977.

Gloster Meteor NF. Mk.14 WS792 'K'

Built 1953, two seat night fighter trainer, two Rolls-Royce Derwent 8s of 3,600 lb s. t.

Built by Armstrong Whitworth at Baginton, WS792 was ready for collection by the RAF on March 8th 1954. It started a long career with Maintenance Units three days later when it was issued to No.8 MU at Little Rissington, followed by No.33 at Lyneham on January 10th 1957, No.27 at Shawbury on January 2nd 1958, No.33 again on February 4th 1958 and No.12 at Kirkbride from March 4th 1960. It then joined No.2 Air Navigation School as a navigation trainer, with the unofficial designation 'NF.(T). Mk.14, from April 8th. It moved to No.1 ANS on January 14th 1962 and here it picked up its code 'K'. It suffered a Category 3 (Repair beyond unit capacity) mishap on December 12th 1963 and was repaired on site by a team from that date until the last day of the year, going back on charge with No.1 ANS the following day. It was retired to No.5 MU at Kemble on January 19th 1966 and declared a non-effective airframe on March 8th. It was allocated to No.2 School of Technical Training at Cosford on June 19th 1967 and became 7965M. It was later transferred to the Museum.

Gloster Meteor NF. Mk.14 WS838

Built 1954, two seat night fighter, two Rolls-Royce Derwent 8s of 3,600 lb s. t.

WS838 was also built at Baginton and was ready for collection on April 28th 1954. It was delivered to No.15 Maintenance Unit at Wroughton two days later for store. On May 9th 1955 it became 'X' of the All-Weather Operational Conversion Unit (No.238 OCU) and moved on to No.64 Squadron on January 20th 1958 for its last, and brief, phase of true operational flying. It was flown to No.12 MU, Kirkbride, on April 1st 1958. It was loaned to the Ministry of Supply from July 30th 1959 (officially transferred to the MoS on November 1st 1961) and started eleven years of test work. This was first undertaken with the All Weather Wing, followed by the Royal Radar Establishment at Pershore and finally with the Royal Aircraft Establishment at Thurleigh. It was retired in late 1971 and earmarked for the RAF Museum, total flying time having amounted to 1,900 hours. It was delivered to Colerne in February 1972 in its yellow test scheme. It was roaded to Shawbury on Colerne's closure on September 11th 1975 and was on display at Cosford in time for the Aerospace Museum's first open day of the 1976 season, in April.

Gloster Javelin FAW. Mk.1 XA564

Built 1955, two seat all weather fighter, powered by two Armstrong Siddeley AS.Sa.6 Sapphires of 8,150 pounds static thrust each.

XA564 was one of the many development and trials Javelins. It was handed over from Gloster's on September 30th 1955. It was in use with the Bristol Aeroplane Company from July 12th 1957 for engine development and other trials, based at Filton. It was retired in 1959 and allocated the maintenance serial 7464M at No.5 School of Technical Training at Locking. It was moved to No.2 SoTT at Cosford in 1962 and later joined the Museum.

Handley Page HP.80 Victor B. Mk.1 XA923 'N'

Built in the mid- 1950s, long range strategic bomber, powered by four Bristol Siddeley Sapphire 202s of 13,000 pounds static thrust each.

XA923 was initially delivered to No.10 Squadron, before joining the Central Signals Establishment for electronic warfare and strategic reconnaissance duties. It was posted to No.232 Operational Conversion Unit 'A' Flight and used to train the crews for No.543 Squadron's Valiants. In April 1958 it joined the Radar Reconnaissance Flight. On May 26th 1964 it was retired to No.2 School of Technical Training at Cosford and allocated 7850M. During 1975 it was given the No.2 SoTT code 'N' and then passed on to the Museum in 1976.

Handley Page HP.115 XP841

Built 1960, single seat slender delta wing research aircraft, powered by one Armstrong Siddeley Viper ASV.9 turbojet of 1,900 pounds static thrust.

The beautiful HP.115 was built during 1960 and 1961 at Radlett to undertake research by the Aerodynamics Flight of the Royal Aircraft Establishment at Thurleigh into the flight characteristics of slender delta wings at low speed. It first flew at Thurleigh on August 17th 1961 and was to remain at Thurleigh for the rest of its flying life. It was retired in late 1973 and flown to Colerne on February 1st 1974. It was taken by road to Cosford on October 9th 1975.

Hawker P.1052 VX272

Built 1948, single seat swept wing research aircraft, one Rolls-Royce Nene.

The P.1052 was a direct development of the Sea Hawk, and featured a 40 degree swept back wing. VX272 first flew on November 19th 1948 and was used for development work into possible production versions and into the problems of swept wings in general. It was navalised to some extent and undertook carrier trials on H.M.S. 'Eagle' in 1952. On retirement it was allocated the maintenance serial 7174M & became an early member of the Colerne Museum. On the closure of Colerne it was moved to St. Athan on September 8th 1975 and put into store. It was moved to Cosford in early 1977.

Hawker Hunter F. Mk.1 WT555

Built 1953, single seat fighter, one Rolls-Royce Avon.

WT555 was the first production Hunter F. Mk.1 and first flew on May 16th 1953. It was retained by Hawkers for trials at Dunsfold until November 1953 when it was handed over to the Aeroplane and Armament Experimental Establishment at Boscombe Down for acceptance trials. It was retired on November 21st 1957 when it was allocated 7499M at No.5 School of Technical Training at Locking for technical instruction. It was moved to Cosford in 1967.

Hunting H.126 XN714

see photo 7

Built 1963, single seat low speed jet flap research aircraft, powered by one Bristol Siddeley Orpheus turbojet exhausting through the wings and tail nozzles.

The odd-looking H.126 first flew on March 26th 1963 and was used mostly by the Aerodynamics Flight of the Royal Aircraft Establishment at Thurleigh for trials into the low speed handling of aircraft. It appeared at the Paris Salon at Le Bourget in June 1965. It was taken to Holme-on-Spalding Moor in 1969 and dismantled prior to being flown to the United States for use with the National Aeronautical and Space Administration. It was flown out to the Ames Flight Research Centre at Moffett Naval Air Station, California from Holme on April 3rd 1969 on board Belfast XR366. It returned to Thurleigh in 1970 and was not unpacked from its crate. It was offered for disposal in June 1972. It was allocated to the Cosford Aerospace Museum on April 30th 1974.

Kawasaki Type 5 Model 1b (Ki 100) BAPC No.83 '24'

Built 1945, single seat fighter, one HA.112-71 radial piston engine.

This aircraft is known to have been built as a Ki 61-II 'Tony' with a HA.140 in-line piston engine, but to have been converted at an early stage (probably on the line) to the radial engined variant. Its service use is unknown. It was captured in 1945 and imported into the United Kingdom in 1946. It is displayed in the colours of the 5th Fighter Squadron, which was based at Chefu and Yakkaichi. Since its acquisition by the Cosford Aerospace Museum it has been allocated the British Aircraft Preservation Council identity number 83.

Messerschmitt Me 163B-1 Komet 191614

Built 1943/4, single seat interceptor, powered by one Walter 109.509 bi-fuel rocket motor.

Captured in Germany in 1945 and transferred to the Royal Aircraft Establishment at Farnborough for evaluation, almost certainly with a so-far untraced 'AIR MIN' number. It was later moved to the Rocket Propulsion Establishment at Westcott (itself a division of the RAE) and kept there for many years, latterly on display with the more recent British equivalent rocket-powered prototype, the SR-53. It was moved to Cosford in 1975.

Messerschmitt Me 410A-1/U2 Hornisse 470430

Built 1943, light bomber, two Daimler-Benz DB 603A piston engines of 1,750 h.p. each.

The identity quoted for this aircraft is thought most likely, although it could also be 410478. It was built in 1943 at Augsburg and is known to have been coded 'PD+VO' at one stage. It served with the 26th Heavy Fighter Wing and saw service in the Balkans and Italy. It was captured intact on the ground in April 1944. It was brought to the United Kingdom and evaluated by the Royal Aircraft Establishment at Farnborough and then with the Air Fighting Development Unit at Wittering and then Ford. The aircraft could also be confused with 10259/TF209 however! 470430 was allocated AIR MIN 72 and was stored at Fulbeck, Gaydon, Rufforth and Biggin Hill before joining Cosford. It was loaned to the RAF Museum for the 'Wings of the Eagle' Display at Hendon in 1975.

Short SB.5 WG768

see photo 48

Built 1952, constructor's number SH.1605, single seat Lightning aerodynamic development aircraft, powered by one Bristol Siddeley Orpheus turbojet.

The Short SB.5 was built to Specification ER.100 as an aircraft capable of finding the optimum format for the planned English Electric P.1 interceptor (later named the Lightning). It could fly with different amount of wing sweep (this could not be changed in flight) and with a conventional or a 'T' tail. (The rear fuselage with 'T' tail is currently stored with the RAF Museum at Henlow, the aircraft being displayed with the conventional tail rear fuselage.) WG768 first flew on December 2nd 1952 from Boscombe Down, having been built at Sydenham. For its initial flights it was fitted with a Rolls-Royce Derwent. Its first tests were undertaken with a 50 degree sweep. On July 29th 1953 the SB.5 first flew with 60 degree sweep and was taken out of service in 1957 for the change to the Orpheus and to 69 degree sweep. It first flew in this guise on October 18th 1960. It was later transferred to Boscombe Down were it joined the Empire Test Pilots School, coded '28' and wearing a dark blue colour scheme. It was retired by 1968 and allocated 8005M and displayed at Finningley, later moving on to Topcliffe and Cosford.

Supermarine Spitfire FR. Mk.XIVe MT847

Built 1945, single seat fighter, one Rolls-Royce Griffon 65 piston engine of 2,050 h.p.

MT847 was first delivered to No.6 Maintenance Unit at Brize Norton on February 28th 1945. On December 15th 1945 it was issued to the Aeroplane and Armament Experimental Establishment at Boscombe Down for tests, returning to store, this time at No.29 MU High Ercall on February 18th 1946. It emerged again on November 14th 1950 when it was taken on charge with

No.226 Operational Conversion Unit at Stradishall. Here it was coded 'UU-A'. It went then to No.33 MU at Lyneham and declared a non-effective airframe on February 9th 1952. It moved on to No.90 MU at Warton from April 30th. Here it was displayed on the gate and was allocated the maintenance number 6960M on May 28th. It moved only a short distance to the gate of Freckleton, and here gained the 'code' 'MT-E' which reflected Freckleton's role as the Medical Training Establishment. It then may have moved to Middleton St. George but this cannot be confirmed. It then was to be found at Weeton before moving to Cosford in 1964 to join the collection.

Supermarine 510 VV106

Built 1948, single seat swept wing development aircraft, one Rolls-Royce Nene turbojet.

VV106 was a cross-breed between the Attacker naval jet and the theories of swept wing aerodynamics. The 510 was not intended for production and was a test vehicle which eventually gave rise to the Supermarine 535, basically a tricycle undercarriage version of the 510, and to the famous Swift. VV106 first flew on December 28th 1948 and was used throughout its operational life for test purposes. It did have a brief spell of stardom when it appeared in the film 'Sound Barrier'. It was allocated 7175M and was an early exhibit at the Colerne Museum. For many years it was displayed in a bare metal scheme with no markings at all, but its refurbishment was more or less complete by the time Colerne closed down. It was moved by road to St. Athan on September 29th 1975, but was moved on to Cosford in early 1977.

Vickers 668 Varsity T. Mk.1 WF408

Built 1952, constructor's number 554, multi-purpose crew trainer, powered by two Bristol Hercules 264 radial piston engines of 1,950 horse power each.

WF408 was built at Hurn and first flew from there on February 29th 1952, joining No.201 Advanced Flying School on March 5th. Here it was coded 'J'. After this it went on to No.11 Flying Training School, No.1 Radio School and No.2 Air Navigation School. With the latter WF408 became 'R'. By 1970 it was on charge with No.6 Support Squadron, based at Cosford and used for communications duties. It was retired at Cosford and officially struck off RAF charge on November 12th 1973, becoming 8395M with the resident No.2 School of Technical Training. It joined the Museum collection in early 1976.

Vickers 701 Viscount G-AMOG

Built 1952, constructor's number 7, medium range airliner, powered by four Rolls-Royce Dart R.Da.3 Mark 506 turboprops of 1,547 estimated shaft horse power each.

G-AMOG first flew from its birthplace at Weybridge on February 11th 1953 and was delivered to British European Airways at Heathrow on March 27th. With BEA it was christened 'Robert Falcon Scott' and was to help in the pioneering of many turboprop routes into Europe and throughout the United Kingdom. It was sold to Cambrian Airways at Rhoose on December 28th 1962, the Viscount 700 having been replaced in BEA service with more Viscounts - this time the larger capacity 800 series. It served with Cambrian in a variety of colour schemes faithfully until April 1972. From that date it was leased to British Overseas Airways Corporation (shortly to become British Airways) and painted in their dark blue and white hue. It was named 'Scottish Prince' and was used for shuttle services based on Prestwick. It was retired in early 1976, and flown from Prestwick to Rhoose on March 31st for what seemed to be an almost certainly violent fate at the wrong end of the scrap-merchant's axe. However it was acquired by the Royal Air Force Museum and flown into Cosford on April 17th where it is displayed in British Airways colours.

DUXFORD

Duxford Airfield, near Cambridge, now housing one of the largest collections of historic aircraft in the United Kingdom has evolved in its present form from a unique experiment of a major Museum opening its doors to outside enthusiast help.

The types preserved at Duxford are both military and civil and range from the extremely valuable Imperial War Museum RE.8 to Concorde 01 recently donated to the Duxford Aviation Society, and due to arrive in early 1977. D.A.S. members are working to restore and preserve aircraft belonging to the Imperial War Museum and the Shuttleworth Trust as well aircraft owned by the Society and its members.

Many aircraft are subject of major restoration projects including six further types for flying condition, adding to the dozen plus currently listed as airworthy.

The Museum works in very close co-operation with the Society on all matters concerned with running the site, the climax of all work each year being the annual flying displays attracting many visiting aircraft.

Duxford airfield is open on March 5th/6th, 12th/13th, 19th/20th and then daily from March 26th, opening hours 1100 to 1700 hours.

Admission is free, car parking charges to be announced.

Address : The Imperial War Museum, London Office, Lambeth Road, London SE1 6HZ.

Telephone : 01 735 8922. (Note : The Imperial War Museum aeronautical collection at Lambeth is presented seperately under the heading Imperial War Museum.)

Address : Duxford Aviation Society, Duxford Airfield, Duxford, Cambridge CB2 4QR.

Telephone : 0223 834312.

Compilers' Note : Duxford is unique in that aircraft under active restoration are mostly open to public inspection, as such aircraft in this condition are listed in the full manner, and not as a footnote in the case of other entries in this book. There are several aircraft at Duxford that may be transitory, or are not complete airframes, these have been appended in briefer format.

Auster B5 AOP. Mk.9 XP281

Built 1961, constructor's number AUS/B5/167 (unconfirmed), aerial observation post, powered by one Blackburn Bombadier 203 of 173 horse power.

XP281 was completed and ready for collection on May 11th 1961, but its first positive use in service is known to be with the Advanced Fixed Wing Flight at the Army Air Corps Centre, Middle Wallop from about August 1966 onwards. It was still to be seen at Middle Wallop during March 1970. On March 10th 1970 it was delivered to the Imperial War Museum at Duxford as a gift from the Army Air Corps.

Avro 652A Anson C. Mk.19 Phase 2 TX183

see photo 15

Built 1945, light communications aircraft, two Armstrong Siddeley Cheetah 15s of 420 h.p.

TX183 is first recorded as serving with the BG Communications Flight (BG?) from February 19th 1946. It went back to the builders from September 26th 1949 until November 30th moving to No.22 Maintenance Unit at Silloth the following day. On February 6th 1950 it was on charge at Abingdon, possibly with the Station Flight. From April 5th 1951 it was in service with No.1 Air Navigation School, and suffered a Category 3 (Repair beyond unit capacity) only four days later. An Avro party repaired TX183 on site from April 17th to (officially) May 2nd. It was officially back on strength with No.1 ANS on May 1st. TX183 flew on unscathed until June 23rd 1952 before being inflicted again with a Catgory 3 blow. No.49 MU at Faygate took charge of this repair, from June 30th to September 10th. It was back with No.1 ANS the following day. It moved on to join the Central Navigation and Controller School at Shawbury from December 19th 1953 and served without major mishap until December 30th 1957 when it returned to No.22 MU. It moved on to No.20 MU at Aston Down on June 3rd 1958, returning to No.22 MU on August 5th 1958. It renewed its acquaintance with No.49 MU on March 23rd 1959 for 'special fitments', these being completed on June 27th. TX183 then found itself back at No.22 MU from July 13th and at No.20 MU from October 29th. April 6th 1960 saw TX183 back into service, this time with the Handling Squadron only to shortly gain its third Category 3 accident! No.71 MU from Bicester repaired the aircraft from January 19th 1961 and it was back with the HS on February 9th. There now was to follow a hectic (and damaging) phase to poor TX183's career. Back with the HS TX183 again went Category 3, and was repaired on site by No.71 MU from June 19th 1962 until July 10th. Back on charge it again sustained Category 3 damage and No.71 MU again despatched a team to go and tend to TX183's needs, from September 26th until October 31st 1962. Back on charge officially on November 1st TX183 resumed its flying duties, and again sustained Category 3 bruises. No.71 MU had by this time lost all patience with the 'Annie' and a team from No.60 MU at Leconfield dealt with the aircraft, from March 3rd 1966 until May 3rd. Three days later TX183 was back on roll with the HS and survived without major incident to be officially struck off charge on March 6th 1968. It was acquired by the Shuttleworth Trust and flown from the Handling Squadron at Boscombe Down into Old Warden in 1968. Since then it was displayed statically until being dismantled and taken to Duxford for complete restoration in 1973.

Avro 716 Shackleton MR.3 Phase 3 XF708 'C'

see photo 25

Built 1958, long range maritime patrol aircraft, powered by four Rolls-Royce Griffon 57A piston engines of 2,450 h.p. each and fitted with two Viper turbojets for take off boost.

XF708 was built as an un-boosted MR.3 and first served with No.201 Squadron, coded 'O' from 1959. It then joined No.120 Squadron, this time as 'A', from 1964. It then returned to the makers at Woodford for the fitting of the Viper turbojets into the rear of the in-board engine nacelles and other detail modifications to bring it up to Phase 3 standard. It then joined its third unit, No.203, and was coded 'C'. It is still displayed in these colours. In 1972 it was flown into No.5 Maintenance Unit at Kemble for storage. It was acquired by the Imperial War Museum and flown into Duxford on August 22nd 1972.

Avro Canada CF-100 Mk.IV Canuck 18393

Built 1955, two seat all weather fighter, two Orenda 11 turbojets of 7,272 lb s. t. each.

18393 was taken on charge by the Royal Canadian Air Force on May 5th 1955 and served with No.440 Squadron, part of the 4th Fighter Wing, for most of its operational life. It was struck off charge on October 3rd 1962 officially. It is thought to have been left by the RCAF at Cranfield (where the type had been based) and the airframe was taken on by the College of Aeronautics there and used as an instructional airframe. It was purchased by Ormond Haydon-Baillie in 1975 and prepared for a ferry flight to Duxford. For this the civil registration G-BCYK was allocated, though not worn on the aircraft. 18393 was flown Cranfield-Duxford on March 29th 1975 and restoration work to render the aircraft totally airworthy has continued ever since.

B.A.C. Drone G-AEDB

Originally built 1936, c/no.13, single seat ultra-light, one Bristol Cherub III of 32 h.p.

Drone G-AEDB as it stands now has had a very involved history, and no better person than its restorer, Bob Ogden, could supply its history. It was completed at Hanworth in March 1936 with a Douglas Sprite of 25 h.p., it was flown to Scone soon after by its first owner, G. Scott Pearce. In October 1936 it was sold to the Ely Aero Club, and later operated by Cambridgeshire Flying Services. In 1937 an in-flight engine failure during a flying display brought G-AEDB down in a forced landing, which destroyed the nose section. It lay unused at Ely until late that year when Humphry Dimock (who owned both EAC and CFS) rebuilt it with a Bristol Cherub III engine and tricycle undercarriage. CFS closed down in late 1938 and G-AEDB was sold to Don Burgoyne of Solihull. Here it was reconverted to normal configuration and consumed parts of Drone G-AEEN (which crashed on August 24th 1937) in the process. The conversion was finished in 1939 but the war prevented flight (actually four hops) until March 31st 1946 from Hockley Heath. Following flights were more successful and G-AEDB was flown cross-country. It was sold briefly to Percy Nadin of Stoke-on-Trent but soon repurchased by Don and stored. Don Graham then bought it and took it by road to Aspatria, where he intended renovating it to flying condition. Mr. Graham lent the Drone for display in Carlisle Market Hall. After display the group who had borrowed it stored the Drone on a farm, only to have the wings trampled to destruction by the local cow population. In March 1968 Bob Ogden took a hand and bought the remains and removed them to Solihull. In early 1970 the wings of Drone G-AEJH (dismantled in Birmingham September 1948) were acquired from Thetford. In that summer the rebuild moved, with Bob, to Woodley. A tremendous amount of work began to show forth by May 1973 when the Drone was ready for final assembly. The Drone was moved to Blackbushe and taxying trials were made in July 1973 (these included a hop of about 150 yards). In late 1973 the Drone returned to Woodley. In January of the following year G-AEDB moved to Duxford. Demands from other aircraft in Bob's care have meant that the Drone has yet to properly fly again, but this is anticipated to take place during 1977.

BAC/Aerospatiale Concorde 01 G-AXDN

Built 1969-71, constructor's number 13522, production number 01, pre-production development super-sonic airliner, powered by four Rolls-Royce/SNECMA Olympus 593 turbojets of 35,080 lst.

Concorde 01 was presented to the Duxford Aviation Society, for preservation in 1976, and is due to fly into Duxford in early 1977. G-AXDN was the first pre-production model Concorde, and the second to be assembled at Filton. It was registered as G-AXDN in April 1969 and flew for the first time at Filton on December 17th 1971. It has been used extensively on the rigorous testing programme designed to make the Concorde the most-tested and tried aircraft ever to enter service. Operating from the test airfield at Fairford, G-AXDN returned to Filton in 1975 for preparation for withdrawal from service, its task completed and its airframe being uneconomic to convert to production status.

Beagle B.206X G-ARRM

see photo 3

Built 1961, constructor's number B.001, prototype general purpose twin, powered by two Continental IO-470D piston engines of 260 h.p. each (not installed on the aircraft).

G-ARRM was the prototype of the relatively successful 206 series of aircraft built at Shoreham and Rearsby. It first flew from Shoreham on August 15th 1961 and was used for development trials and the testing of production systems and equipment. The Beagle 206X looked different in many detailed features to the production versions. It was withdrawn from use at Shoreham on December 28th 1964 and placed in a protective cocoon in July 1966. Upon the collapse of Beagle, it appeared that this aircraft would be scrapped, but it was saved and presented to the Imperial War Museum for safe-keeping at Duxford, albeit in an engineless state. It was moved to Duxford on August 11th 1970.

Beagle B.121C Pup 100 G-AVDF

Built 1967, constructor's number B.001, prototype light aircraft, one Lycoming flat four piston engine of 200 h.p. in final form - not fitted to the aircraft.

G-AVDF was the prototype Pup and flew from the first time from Shoreham on April 8th 1967 in the hands of J. W. C. Judge, and powered by Rolls-Royce Continental 0-200-A of 100 h.p. Like the prototype 206, above, it was used for development flying. G-AVDF ending its flying days testing out some of the features of the proposed Bulldog military trainer, this being the reason for the 200 h.p. Lycoming installation. It was dismantled on the collapse of Beagle, but was saved and moved to Duxford for the Imperial War Museum in September 1970. Like the 206, it did not have an engine installed when it was acquired.

Boeing B-17G-DL Fortress F-BDRS

see photo 12

Built 1945, constructor's number 32376, photographic survey aircraft, powered by four Wright Cyclone R-1820 radial piston engines of 1,600 horse power each.

F-BDRS was built by Douglas at their Long Beach plant as a standard B-17G bomber on order for the United States Army Air Force, serial number 44-83835. Its service life with the USAAF is not known. Immediately post-war it was sold off as surplus and came on to the American civil register as N68629. It was then extensively internally and externally modified for its new role, that of a long range executive transport. Modifications included the removal of all the gun turrets, the fairing over of the rear gun position, the installation of much navigation and radio equipment, and the placing of a large cabin in the rear fuselage with four windows on either side and a large self-contained air-stair on the starboard side. It flew in this guise for some time, until being sold to the Institut Geographique National as F-BDRS for use as a survey aircraft. It was based at Creil, near Paris, for this task and was to serve until being withdrawn from use in 1972. It remained at Creil intact as a source of spares for the remaining airworthy B-17s of the IGN's fleet. F-BDRS was acquired by Euroworld and donated to the Imperial War Museum in 1975. In mid-1975 a DAS crew went out to dismantle the aircraft to bring it to Duxford. The IGN crews painted a suitably appealing female on the side and gave F-BDRS the name 'La Duxford Belle'. F-BDRS is currently under active restoration to full wartime B-17G standards, for static display.

Boeing B-17G-105-VE Fortress G-BEDF '485784'

photo 13

Built 1945, constructor's number 8693, photographic survey aircraft, powered as F-BDRS.

Duxford's famous airworthy B-17G is owned by Euroworld Ltd and is based at Duxford as a memorial to the men of the United States 8th Air Force who flew and fought from the United Kingdom. It flies in ficticious markings, as 'Sally B'. The aircraft was built by the Vega Aircraft Corporation in 1945 to United States Army Air Force order as a B-17G with the serial number 44-85784. Its service with the USAAF is not known. It was acquired by the Institut Geographique National at Creil in the early 1950s and registered as F-BGSR. It was modified for survey duties and remained airworthy with the IGN until acquired by Euroworld in late 1974. It was flown to Duxford on March 16th 1975 with the American civil registration N17TE. It performed at several displays during the 1975 season. It was maintained on the U.S. register as Euroworld intended flying N17TE to the U.S.A. after a season with Duxford. However the company decided in 1976 to put the aircraft onto the British register and to keep it in the United Kingdom. It was registered G-BEDF in late 1976 for this purpose.

Bristol Bolingbroke Mk. IVTs 9896 and 10038

Both built 1942, target tugs, powered by two Bristol Mercury XVs of 920 h.p. each.

These two aircraft are the property of Ormond Haydon-Baillie who imported them from Canada via Harwich docks in 1974, along with many spare components. They are under active restoration to full flying condition. The Bolingbroke is the Canadian equivalent of the Blenheim IV. Both aircraft were built by Fairchild Aircraft Ltd at Longueil, Port Quebec. 9896 (previously thought to be 9893) was taken on charge by the Royal Canadian Air Force on June 10th 1942 and served with No.8 Bombing and Gunnery School at Lethbridge. It was struck off charge on May 22nd 1946. 10038 was taken on charge on December 6th 1942 and served with No.3 Bombing and Gunnery School at McDonald in Manitoba. It was struck off charge on May 15th 1946, at which time it had a total flying time of 489 hours 35 minutes. Both aircraft became the property of Wes Agnew and were stored in the open for many years until being acquired by Mr. Haydon-Baillie.

Bristol B.175 Britannia 312 G-AOVT

see photo 18

Built 1958, constructor's number 13427, long range airliner, powered by four Bristol Proteus 755 turboprops of 4,120 estimated shaft horse power each.

G-AOVT first flew from Filton on December 17th 1958 and gained its Certificate of Airworthiness on the 30th. It was delivered to its owners, British Overseas Airways Corporation at Heathrow on the first day of 1959 and served the Corporation for just over four years. It was bought by British Eagle International Airlines Ltd., in September 1963 and given the name 'Ajax'. It served with Eagle on that airlines many inclusive tour flights until the tragic collapse of the airline. G-AOVT was then bought up by Monarch Airlines of Luton and used by them for I.T. work and later trooping and freight flights. Monarch took delivery of G-AOVT on May 9th 1969. With the introduction of Boeing 720s and BAC 111s Monarch found that G-AOVT was surplus and donated the aircraft to the DAS, the aircraft today being technically under the joint ownership of the airline and the DAS. On June 28th 1975 this 'Whispering Giant' flew into Duxford. Since then it has been under internal restoration to bring it back to passenger configuration.

Cessna 195 G-BBYE

Built 1950, constructor's number 7550, four seat light aircraft, powered by one Jacobs R-915 7 radial piston engine of 180 horse power.

This aircraft was built for civilian use and was first registered in the United States as N9857A. It was sold in Norway in the early 1960s and registered there as LN-BDV. It was then bought by the Moorsele Paraclub of Belgium, who used it as a platform for parachute dropping. It was registered in Belgium as OO-PCM. It was imported into the United Kingdom in late 1974 by Cliff Lovell and registered as G-BBYE. It is now in the ownership of R. Willies and is one of the airworthy aircraft based at Duxford.

Convair VT-29B-CO 17899

Built 1951, constructor's number 311, staff transport, powered by two Pratt and Whitney R-2800-99W piston engines of 2,500 horse power each.

Built at Convair's San Diego plant for the United States Air Force as a T-29B crew trainer, full serial 51-7899, this aircraft was soon fitted up with an executive interior and put to work as a VIP transport and general communications aircraft. It was based for some time at Northolt, and latterly with the 513th Tactical Air Wing, based at Mildenhall. It was presented to the Imperial War Museum by the USAF on the retirement of the type from the 513th TAW, the type having been replaced by T-39 Sabreliners. It flew into Duxford in early 1975 and is in superb condition.

De Havilland DH.82A Tiger Moth NM181/G-AZGZ

Built 1944, constructor's number 86489, basic trainer, one De Havilland Gipsy Major in-line piston engine of 130 horse power.

Built for the Royal Air Force by Morris Motors as NM181. It was sold in the early 1950s to France and registered as F-BGCF and used by several flying clubs, including the Aero Club de Haute-Moselle. It was imported into the United Kingdom again in 1972 and registered as G-AZGZ. It was painted up back into its wartime livery, with its original serial NM181. It is currently airworthy, and owned by F. Manning.

De Havilland DH.82A Tiger Moth G-APMM

Built 1942, constructor's number 85427, basic trainer, powerplant as for NM181.

Like NM181 above, G-APMM started out in the hands of Morris Motors, for the RAF, this time as DE419. It served with No.21 Elementary Flying Training School from 1942 until 1946, it was coded 'FIVC'. It was registered as G-APMM in September 1953 and went through a series of different owners. In 1962 it started a lengthy career with the Bustard Flying Group and was loaned to the Historic Aircraft Museum for display from 1972, by the Group. In 1976 it was bought by Captain M. C. Russell and it was brought to Duxford late that year. It is to be restored to flying condition and to be used as a tug for Captain Russell's fleet of vintage gliders, also based at Duxford.

De Havilland DH.89A Dragon Rapide G-AGJG

Built 1941, constructor's number 6517, general purpose transport, powered by two De Havilland Gipsy Queen 3 in-line piston engines of 200 h.p. each.

Built by Brush Coachworks Ltd at Loughborough as a Dominie I navigation trainer, serial X7344. It was not fully fitted out to Dominie status and was registered in 1943 as G-AGJG with Scottish Airways until February 1947, when it joined BEA for a year. Since then it has a string of owners and latterly was a well-known parachute club mount operating out of Halfpenny Green airfield. In 1975 it was at Biggin Hill, and became the property of Mr. E. Wein of Express Aviation. On August 17th 1975 G-AGJG touched down at Duxford to start a complete restoration programme, it now being owned jointly by the DAS and Mr. Wein.

De Havilland DH.100 Vampire FB. Mk.5 VZ304

Built 1950, single seat fighter, one De Havilland Goblin turbojet.

VZ304 was ready and waiting for collection by the RAF on February 27th 1950. It was moved to No.33 Maintenance Unit by March 10th and was allocated almost immediately to the Middle East Air Force. It flew to Manston on May 5th and departed the country nine days later, arriving on the 21st. On the 24th it was on charge with No.107 MU, prior to issue to No.249 Squadron on August 18th. It suffered a Category 3 (Repair beyond unit's capability) smash on February 16th 1951, repairs being made by No.109 MU. It was back on strength with No.249 Squadron on March 6th. It arrived back in the United Kingdom on May 12th 1952 and was placed in store with No.48 MU at Hawarden seven days later. On March 31st 1954 it was issued to No.3 Civilian Anti-Aircraft Co-operation Unit at Exeter. It was bent again (again Category 3) and repaired on site by No.49 MU from Colerne between February 3rd and March 18th 1959. February 16th 1960 it was issued to No.22 MU at Silloth and declared a non-effective airframe. Three days later it was allocated the maintenance airframe number 7630M. On April 6th it went to No.14 MU at Carlisle, and was officially struck off RAF charge. Here it was displayed near the main guardroom, and was to remain so until 1976 when it was put up for disposal. It was bought by Mr. A. Topen and removed by road to Duxford in September 1976. It is now under complete restoration to static standard.

De Havilland DH.104 Dove 6 G-ALFU

Built 1948, constructor's number 04234, calibration aircraft, powered by two De Havilland Gipsy Queen 70-2 in-line piston engines of 380 horse power each.

Built at Hawarden as a Series 2 G-ALFU was registered on December 14th 1948. It has served for all its operational life as a navigation aid checker and general transport for the various departments of the government that have been charged with this task. It was officially handed over to the Ministry of Civil Aviation on January 18th 1949. As Dove development went on, so G-ALFU was up-graded to keep pace. It becoming a Series 2B and finally a Series 6. Its next operator was the Minister of Transport and Civil Aviation from December 12th 1953, followed by the Board of Trade from November 15th 1956, the Department of Trade and Industry from January 20th 1971 and finally the Civil Aviation Authority from April 1972. In all this time it was based at Stansted. G-ALFU was officially withdrawn from use on November 14th 1972 and presented by the CAA to the Imperial War Museum, it flying in during early 1973.

De Havilland DH.106 Comet 2R XK695

Built 1956, constructor's number 06030, strategic reconnaissance aircraft, powered by four Rolls-Royce Avon 117 turbojets of 7,300 pounds static thrust each.

Provisionally allocated G-AMXH, XK695 flew for the first time from Hatfield on August 21st 1956 and was delivered to No.216 Squadron as a C. Mk.2 transport on September 14th 1956. It was named 'Perseus' with the unit. On March 15th 1960 it sustained a Category 3 (Repair beyond unit's capacity) flying accident and was repaired by a team from De Havillands and No.49 Maintenance Unit, Colerne. On June 17th it was transferred to Ministry of Aviation charge at Hatfield and its conversion to a reconnaissance aircraft began. On July 31st the following year it passed on to Marshalls of Cambridge at Teversham. The work finally completed, XK695 was handed over to No.51 Squadron at Watton on March 8th 1963. It suffered another Category 3 smash on January 6th the following year, the repair being effected by No.71 MU on site, the team handing the aircraft back to No.51 Squadron on September 21st. On November 8th 1971 it went to Hawker Siddeley for a major overhaul and was then flown to No.23 MU at Aldergrove on April 28th 1972 for storage. It re-joined No.51 Squadron, now at Wyton, on June 12th as problems in introducing the Comet's replacement, the Nimrod R. Mk.1, had slowed the latter's full introduction into service. It was placed in open store at Wyton from 1973 and was later donated to the Imperial War Museum. It flew into Duxford on

10 *Duxford-based Sea Fury WH589.* *Stuart Howe*

11 *Spitfire G-15-109, Old Warden 12/9/53* *R. A. Scholefield*

12 *F-BDRS at Creil 15/6/63 note windows.* *P. H. Butler*

13 *F-BGSR awaits engine, Creil 15/6/63* *P. H. Butler*

14 *Sunderland ML796 in June 1976, D.A.S. crew behind.* *DAS.*

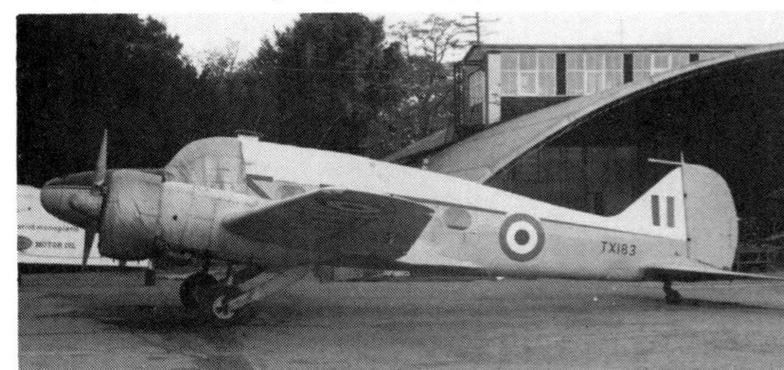

15 *Anson C. Mk.19 TX183 during its days at Old Warden.*

16 *Lightning XM135 taxies out at Coltishall* Roger Lindsay

17 *Javelin XH897 at Coltishall 14/9/68* Roger Lindsay

18 *Britannia G-AOVT with British Eagle at Speke*

19 *Viscount G-ALWF being dismantled, Speke 1976* Ken Ellis

20 *Firefly TT.4 VH127 at Hurn 2/9/57* P. H. Butler

21 *N1854 Fulmar at Farnborough, circa 1963* P. H. Butler

January 10th 1975 and was officially transferred to the charge of the IWM three days later.

De Havilland DH.106 Comet 4 G-APDB

Built 1958, constructor's number 06403, long range airliner, powered by four Rolls-Royce Avon 525B turbojets of 10,500 pounds static thrust each.

G-APDB was built at Hatfield and flew for the first time on July 27th 1958. After initial handling trials it was handed over to British Overseas Airways Corporation on September 30th. On October 4th it flew the first ever trans-Atlantic passenger service by a jet aircraft flying Heathrow-New York. It served with BOAC until June 1965 when it was sold to Malaysian Airways, who registered it as 9M-AOB. It came home in October 1969 and became the property of the most prolific Comet operator, Dan-Air, based at Gatwick. It flew on inclusive tour duties with Dan-Air until being retired to Lasham in 1973. It was presented to the DAS by the airline and flown into Duxford on February 12th 1974. At that time it had amassed a total of 36,269 accident-free hours in the air and had completed its 15,733rd landing.

De Havilland DH.110 Sea Vixen FAW. Mk.2 XS576 '125/E'

Built 1964, two seat all weather naval fighter, powered by two Rolls-Royce Avon 208 turbojets of 10,000 pounds static thrust each.

Built at Hawarden, XS576 was first delivered to the Fleet Air Arm on October 22nd 1964, almost certainly to the Aircraft Holding Unit at Brawdy. By the end of that year it was in use with No.899 Squadron, coded '490/E' and later '125/E' from July 1965 until April 1969. During this time it was embarked on H.M.S. 'Eagle' for operations. On April 2nd 1969 it was delivered to the Royal Naval Aircraft Yard at Sydenham and put into storage. It was officially handed over to the Imperial War Museum there on November 14th 1972 having flown into Duxford on March 17th 1972.

De Havilland DH.112 Sea Venom FAW. Mk.21 XG613

Built 1956, two seat all weather naval fighter, one De Havilland Ghost 104 turbojet.

Taken on charge with the Fleet Air Arm on February 16th 1956, XG613 remained with the makers until at least July 1957 on various trials. It was to be found at Royal Naval Air Station Ford in July 1958. It was transferred to the Mediterranean on October 8th 1958, returning on January 13th 1959. It was at Fleetlands by July 1959 and had joined No.700 Squadron at Yeovilton by January 1960. Here it was coded '519/VL' and remained in service only until that July. It was in store at Abbotsinch from January 1961 until the following January at least. It was struck off charge on December 19th 1962 as sold to the Imperial War Museum. It was displayed on loan to the Shuttleworth Trust in the open at Old Warden for many years, but moved to Duxford in 1971.

De Havilland DH.115 Vampire T. Mk.11 WZ590

Built 1952, constructor's number 15165, advanced trainer, powered by one De Havilland Goblin 35 turbojet of 3,500 pounds static thrust.

The 75th Vampire T. Mk.11 to be built by De Havilland's at Hawarden, WZ590 was first delivered to No.228 Operational Conversion Unit. Nothing more is known until it joined No.5 Flying Training School at Oakington, on charge with that unit's No.3 Squadron, and alloted the code '40'. It then moved to No.8 FTS at Swinderby, and was coded '19'. It was sold to Hawker Siddeley Aviation on December 9th 1968 and put into store at Hawarden, pending possible resale. In 1970/71 it was moved to Woodford. In 1973 it was offered, along with many other Vampires, to the British Aircraft Preservation Council and was acquired from the BAPC by the Imperial War Museum. It was delivered to Duxford by road on May 4th 1973. Since then it has been undergoing a thorough refurbishing.

De Havilland DH.115 Sea Vampire T. Mk.22 XG743 '798/BY'

Built 1955, advanced trainer, engine as for WZ590.

This aircraft was first in use at Royal Naval Air Station Ford with No.764 Squadron, coded '255/FD' from 1955. By June or July of the following year it had been recoded '694/FD'. It then moved to Lossiemouth for service with No.736 Squadron from June 1957, as '597/LM'. XG743's last flying was done with the Station Flight of Brawdy from June 1970 until October 1973. Here it was coded '798/BY' and was used as a communications 'hack', continuation trainer and any other job it could feasibly carry out! It was donated to the Imperial War Museum soon after and flown to Duxford.

De Havilland Canada DHC-1 Chipmunk G-BCYJ

Built mid-1950s, constructor's number C1/0360, primary trainer, powered by one De Havilland Gipsy Major 8 of 145 horse power.

Built for the RAF as WG307 this aircraft last served operationally with Birmingham University Air Squadron from Shawbury as 'U'. It was retired to No.5 Maintenance Unit at Kemble and put up for sale. It was bought by Ormond Haydon-Baillie and registered to him as G-BCYJ. It was flown into Duxford in 1975 and has since been undergoing a major rebuild. It is thought that this aircraft is to remain with Mr. Haydon-Baillie at Duxford, unlike a similar aircraft noted at the end of this listing.

Dewoitine D.26 G-BBMI

Built 1931, constructors number 315, single seat parasol winged fighter, one Jacobs R-755-A2 radial piston engine.

This aircraft was built for the Swiss Air Force, serial number 282, and served with that air arm through the Second World War. It flies today in its original markings. Sometime post-war it was civilianised as HB-RAA and spent most of the remainder of its life in Switzerland at Bern. It was imported into the United Kingdom by 1970 and lay unconverted at Booker. It was registered as G-BBMI in 1974 and brought up to full airworthiness standards. It was based at Old Warden for a while but moved to Duxford in 1975. It is owned by R. Willies and is completely airworthy.

DFS 108-30 Kranich II BGA.1147

Built 1943, constructor's number DFS.821, two seat glider.

This aircraft was acquired by the Royal Air Force Gliding and Soaring Association post-war, initially for use in Germany. It was later imported into the United Kingdom and given the RAFGSA number 215. It was sold after some years of use to Mr. D. Lloyd with whom it was registered with the British Gliding Association as certificate number 1147. It was active as such from November 1963 until November 1966. It was placed into store after this date until acquired by Captain Mike Russell in April 1975 for restoration. It was moved from Bishop's Stortford to Duxford towards the end of 1976.

DFS 108-49 Grunau Baby VN148

Built early 1940s, training glider.

The history of this aircraft was very obscure until it was acquired by the Snuttleworth Trust and then by Mike Russell. Between them, much of its history has been unearthed. This aircraft was captured in 1945 in Germany, were it had been in use with the Luftwaffe with the coding 'LN+ST'. It was imported into the United Kingdom and given the military serial VN148. It was sold to the College of Aeronautics on July 31st 1947 and thence to the Shuttleworth Trust. VN148 was mostly in store at Old Warden. It was allocated the British Aircraft Preservation Council identity number 33 in the absence of any other known identity. By 1975 the Grunau had been overhauled by the Trust and was seen wearing the serial VN148. Early in 1976 it was handed over to Mike Russell for restoration. By late 1976 it had been installed at Duxford and the early German markings were rediscovered.

English Electric Canberra B. Mk.2 WH725

Built mid-1950s, constructor's number 71206, medium bomber, powered by two Rolls-Royce Avon 109s of 7,100 pounds static thrust each.

WH725 first served with No.44 Squadron. It followed this with real action as part of 'Operation Musketeer', the Suez campaign of 1956, with No.50 Squadron. It wears the scheme of those days at present, including the yellow and black 'Suez stripes'. By 1965 it was with the fire section at Yatesbury, but for some reason did not suffer a scorching and was moved to No.15 Maintenance Unit at Wroughton for storage. It was presented to the Imperial War Museum in 1972 and was painted up in its original No.50 Squadron colours.

English Electric Lightning F. Mk.1A XM135

see photo 16

Built 1959, fuselage number 95031, single seat interceptor, powered by two Rolls-Royce Avon 200 turbojets of 10,000 pounds static thrust.

XM135 flew from Warton for the first time on November 14th 1959 and was delivered to the Air Fighting Development Squadron at Wittering on May 14th 1960, with whom it was coded 'D'. It was briefly with the Aeroplane and Armament Experimental Establishment at Boscombe Down from April 1961. It then saw service with the 'Tiger' Squadron, No.74, as 'R' for August and September 1961, before returning to the AFDS. It re-joined No.74 Squadron at Leuchars from September 1962, this time as 'B'. It joined No.226 Operational Conversion Unit on July 25th 1964. It then moved to No.33 Maintenance Unit at Lyneham as the aircraft was experiencing repeated problems with its fuel system. Here it was to take part in a most amazing event. XM135 arrived at Lyneham on January 12th 1965. Shortly after this date the

Commanding Officer decided that the only way to find out what was happening to the fuel flow was to try the aircraft out for himself. He organised a ground running session with himself in the cockpit to monitor the problem as it occured. The Lightning was stripped of most of its cockpit equipment, apart from instrumentation and flying controls, it had no cockpit canopy and no ejector seat installed. During the ground run the throttle gave problems and the supersonic fighter rushed down the runway, with a non-pilot in command!! XM135 took off and with considerable coolness the newly-created pilot managed to bring the aircraft round for a circuit and an almost text-book landing. Cured, XM135 was despatched to No.60 MU at Leconfield on December 12th 1965 for re-allocation to a unit. This happened on September 22nd 1966 when the Lightning again touched the concrete of Leuchars, this time for service with the Target Facilities Flight. It was retired to No.60 MU on June 29th 1971 and was stored until February 1972. On this date it was put onto strength with the MU as a 'hack' and was christened '60 MU Flagship'. It was declared a non-effective airframe on July 31st 1974 and presented to the Imperial War Museum. It was flown into Duxford from Leconfield on November 11th 1974. It had completed 1,343 hours in the air.

EoN AP.5 Olympia

Intended hybrid rebuild of two aircraft, both originally built 1947, single seat glider.

Mike Russell has brought together two Olympias and plans to rebuild them into one fully airworthy aircraft. They arrived at Duxford in late 1976. One aircraft is BGA.537, constructor's number EoN/0/039, which was registered in September 1947, it later took up the registration G-ALKA, but dropped this identity before 1964. It was converted to Mk.2 standards. It crashed at Sutton Bank on October 16th 1970. The second machine was BGA.540, constructor's number EoN/0/042, also registered in September 1947. It later became part of the Royal Air Force Gliding and Soaring Association's fleet, and was given their identity number 145. It returned to civilian use in July 1962 as BGA.1057.

Fairey Type Q Gannet 'AS.6' XG797 '766/BY'

Built 1957, constructor's number F.9365, three seat electronic counter-measures aircraft, powered by one Bristol Siddeley Double Mamba 101 coupled turboprop of 3,025 e.s.h.p.

Built in 1957 at Hayes as an AS. Mk.4 anti-submarine aircraft, XG797 was taken on charge by the Fleet Air Arm on April 4th of that year at Royal Naval Air Station Anthorn. It was at RNAS Abbotsinch by July and was with No.810 Squadron from May 1st 1959 as '232'. It left on H.M.S. 'Centaur' from a tour of the Mediterranean on June 8th and from the 30th on a tour to the Far East, returning home on April 19th 1960. It was back on 'Centaur' again during July 1960. It was with No.700 Squadron from January 1961 until that June, based at Yeovilton, and coded '505/VL'. It was at Abbotsinch from July 1961 until January 1963. It was flown to Culdrose by July 1963 and on to Lee-on-Solent by January 1964. It stayed there for a year and during this time was converted to an electronic counter-measures aircraft, under the designation ECM. Mk.6 (or even E. Mk.6, the above designation appears to be used most often, although it is unofficial.). It was taken on charge with No.831 Squadron at Watton on May 18th 1965. Its use in this new role was strictly limited as it was retired to Arbroath on June 2nd 1966 and used as an instructional airframe. It was acquired by the Imperial War Museum in 1972.

Focke-Achgelis Fa 330A-1 Bachstelze 100143

The type built from 1942, single seat submarine-launched observation gyroglider.

Many examples of this fascinating observation gyroglider were brought to the United Kingdom in 1945. They were built by Weser Flugzeugwerke at Delmenhorst, about two hundred of the production version, the Fa 330A-1, being believed to have been completed. Many were used by the Royal Aircraft Establishment at Farnborough and the Airborne Forces Experimental Unit at Beaulieu. This example 'appeared' at Duxford in the middle of 1976 and it may have been used at Farnborough and/or Beaulieu, but may also have been ear-marked for the Imperial War Museum from the beginning, and have been in store ever since. This example has a full instrument panel, but is missing rotors and rigging.

Gloster Meteor F. Mk.8 WK991

Built 1953, single seat fighter, two Rolls-Royce Derwent 8s of 3,600 pounds static thrust.

WK991 was ready for collection from Hucclecote on July 8th 1953 and was delivered to No.20 Maintenance Unit at Aston Down two days later. On August 10th it joined the Northern Sector Flight at Linton-on-Ouse, and was at one time coded 'VSB' - the initials of Air Vice Marshall V. S. Bowling. It then remained on the 'hack'/communications theme with a move to No.13 Group Communications Flight from November 21st 1957. WK991 joined its first Squadron - No.46 - on March 24th the following year. It served until receiving a Category 3 (Repair beyond unit capacity) accident on June 13th - this requiring the healing treatment of a party from No.71 MU at Bicester, from June 18th until August 21st, returning to the Squadron's official charge the following day. On July 29th 1959 the Meteor stayed with a Squadron, but reverted to its earlier role of 'hack'. This time it was with the Hunter unit No.56 Squadron and was initially coded 'V' and later 'A'. It went to No.5 MU at Kemble on December 20th 1960 and was declared a non effective airframe on April 15th 1961. It became 7825M on November 28th 1963 and was struck off charge on December 10th as acquired by the Imperial War Museum. For some time WK991's nose section was on display at Lambeth, but with the establishment of Duxford, this was reunited with the remainder of the airframe and is still resplendent in its No.56 Squadron colours.

Gloster Meteor NF. Mk.11 WD686

Built 1952, two seat night fighter, two Rolls-Royce Derwent 8s of 3,600 pounds static thrust.

WD686 was built by Armstrong Whitworth at Baginton and was ready for collection on April 25th 1952. It was to serve all its life with test establishments. It was on strength with the Telecommunications Research Establishment at Defford from May 27th but was 'traded in' for WD591 on February 28th 1955. WD686 was at No.15 Maintenance Unit at Wroughton from November 29th 1955. It next moved to the Royal Aircraft Establishment at Thurleigh (on charge officially with the Ministry of Aviation) and was later fitted with a nose-probe. By October 1972 it had been retired and stipped of many components and dumped on the fire heap. It was rescued by the Imperial War Museum and brought to Duxford where its rebuild and refurbishment continues.

Gloster Meteor NF. Mk.14 WS760

Built 1954, two seat night fighter, two Rolls-Royce Derwent 8s of 3,600 pounds static thrust.

Another Baginton product, WS760 was awaiting collection there on February 9th 1954. Thirteen days later it was with Avros for a period of modification, ending on March 11th. Fourteen days later it was in the care of No.15 Maintenance Unit at Wroughton. It joined the All Weather Operational Conversion Unit (No.237 OCU) on May 10th 1955. This was followed by operational service with No.64 Squadron as 'U' from April 2nd 1958 only until September 23rd when it was put into store at No.33 MU, Lyneham. From March 11th 1964 WS760 was at No.5 MU Kemble receiving refurbishment to render it an NF(T). Mk.14 and it was issued to No.1 Air Navigation School on April 16th. Here it became 'P'. It was back at Kemble on January 19th 1966 and declared non-effective on March 8th. It moved to Upwood on August 23rd 1967, having been allocated the maintenance serial 7964M on June 19th. By March 1970 it was installed on the gate at Brampton, the home of Training Command. It was put up for disposal in 1976 and replaced by a Gnat T. Mk.1. It was bought by Mr. A. Topen and brought by road to Duxford on August 8th 1976, it is being refurbished to possible flying status.

Gloster Javelin FAW. Mk.9 XH897

see photo 17

Built 1958, two seat all weather fighter, powered by two Bristol Siddeley Sapphire turbojets of 11,000 pounds static thrust (12,300 lbs s.t. with after-burner) each.

XH897 was built as an FAW. Mk.7 and was taken on charge on December 24th 1958, being with No.19 Maintenance Unit, St. Athan, six days later. It briefly joined No.25 Squadron, as 'M', from March 12th 1959 until April 12th 1960 when it returned to Gloster's for conversion to FAW. Mk.9 standard at Hucclecote. It completed this work on November 4th and was back at St. Athan by the 9th. On January 3rd 1961 it joined No.33 Squadron at Middleton St. George coded 'W'. No.33 became No.5 Squadron on November 21st 1962 and, still as 'W', XH897 moved with the unit to Gilenkirchen in West Germany. It was retired to the grave-yard of many 'Flat Irons', No.27 MU at Shawbury on October 9th 1963. It was spared the axe however and went to Bristol Siddeley Engines for test work at Filton from July 20th 1965. It then moved to the Aeroplane and Armament Experimental Establishment at Boscombe Down on April 1st 1968. It retained its camouflage until at least June 1968 when it gained the superb red and white scheme it now wears, this was reflecting the aircraft's role of a calibration checker for other aircraft's instruments. Upon retirement, replaced by a Phantom, it was acquired by the Imperial War Museum and flown into Duxford on January 24th 1975.

Handley Page HP.80 Victor B(K2P). Mk.1A XH648

Built late 1950s, two point air-to-air tanker, powered by four Bristol Siddeley Sapphire 202 turbojets of 11,000 pounds static thrust each.

XH648 was built at Radlett as a B. Mk.1 strategic bomber, and served in this role first with No.15 and then No.55 Squadrons. It then returned to the makers for conversion to a tanker aircraft, firstly with the designation BK. Mk.1A, but this was later changed to the designation quoted above. It then was delivered to Marham, and its old unit, No.55 Squadron, now flying in a totally different role. It finally served with No.57 Squadron until acquired by the Imperial War Museum and flown into Duxford on June 2nd 1976.

Hawker Sea Fury FB. Mk.11 WH589 'OH-B'

see photo 10

Built 1950, single seat naval fighter bomber, powered by one Bristol Centuarus 18 radial piston engine of 2,480 horse power.

This aircraft was originally delivered to Royal Naval Air Station Anthorn, but soon issued to the Royal Australian Navy, who continued to use it as WH589. It was sold off as surplus at Bankstown in the early 1960s and was acquired by Ormond Haydon-Baillie, then serving with the Royal Canadian Air Force at Cold Lake, Manitoba. It was shipped to Canada and became CF-CHB. It was painted up in RAF-style World War Two colours and given the 'squadron code' 'OH-B'. On completion of Mr. Haydon-Baillie's tour of duty the Sea Fury (and his two T-33s, see below) were ferried to the United Kingdom, the Sea Fury landing at Southend on November 24th 1973. It was registered as G-AGHB on May 9th 1974, but this was not painted on the aircraft. It moved to Duxford after only a short stay at Southend. It remains fully airworthy and is often seen at air displays in the United Kingdom.

Hawker Sea Hawk FB. Mk.5 WM969 'SAH-5'

Built 1954, single seat naval fighter bomber, one Rolls-Royce Nene 103 of 5,400 lbs s.t.

As this book goes to type, this aircraft is the latest arrival at Duxford, arriving by road from Culdrose on January 20th 1977. This aircraft had previously been in use with the School of Aircraft Handling at Culdrose as a ground runable instructional airframe used to train aircraft carrier flight deck marshallers in the skills of parking and directing aircraft. For this it was given the SAH number 5 and the naval maintenance number of A2530. It had been in use at the SAH since at least 1969.

Hawker Hunter F. Mk.2 WN904 '3'

Built 1954, single seat fighter, one Bristol Siddeley Sapphire.

WN904 was built by Armstrong Whitworth Aircraft at Baginton, and flew for the first time on July 19th 1954, going on to join No.257 Squadron as 'Q'. It was retired and was allocated the maintenance serial 7544M at Melksham on December 3rd 1960. It later moved to No.9 School of Technical Training at Newton, gaining the code '3' in 1967. No.9 SoTT was closed down in 1974 and the Hunter was acquired by the Imperial War Museum and moved to Duxford by road on May 11th 1974.

Hawkridge Dagling BGA.493

Built 1947, constructor's number 10471, single seat primary glider.

This aircraft was one of a batch built post-war by the Hawkridge Glider Company and registered with the British Gliding Association as certificate number 493. It was used by the Derby and Lancs Gliding Club and later by the Blackpool and Fylde Gliding Club in the 1950s. At this stage in its life it lost its identity. It was bought by Mr. Ivor Stretch and stored at Warton, and later at Squires Gate with No.177 Squadron, Air Training Corps. It was allocated the British Aircraft Preservation Council identity number 81 in the absence of any other identity. In late 1976 it was loaned to Captain Mike Russell and moved to Duxford to be restored to airworthiness. Early inspection of the airframe brought to life its long-buried identity.

Junkers Ju 52/3mge

Built late 1930s or early 1940s, general purpose transport, powered by three BMW 132Z radial piston engines of 770 horse power each.

Very little is known about this aircraft, it could be a French built AAC-1 variant, supplied by the French Air Force in the mid-1960s. Whatever its original source it was serial number 6316. After many trials and tribulations in the dockyards of Portugal the aircraft was delivered by road to Duxford in May 1974. During the first half of 1976 it was repainted in typical Luftwaffe temperate camouflage of World War Two.

Lockheed T-33s of Ormond Haydon-Baillie

Advanced jet trainer, all three powered by Rolls-Royce Nene turbojets.

Ormond Haydon-Baillie is the owner of a unique collection of aircraft, all of which are based at Duxford, the bulk of his collection are three T-33 'T-birds', as follows :-

T-33A/N G-OAHB The famous black and gold 'Black Knight', star of several TV films and many air shows. It was built by Canadair for the Royal Canadian Air Force as a Silver Star Mk.3, constructor's number 261, serial number 21261. It was bought by Mr. Haydon-Baillie whilst he was with the RCAF at Cold Lake and civilianised as CF-IHB. It was flown across the Atlantic, and arrived at Southend on November 9th 1973, taking up the 'fixed' registration G-OAHB. It moved to Duxford with the rest of the fleet in 1974.

T-33A/N G-WGHB This aircraft was built also as a Silver Star Mk.3, constructor's number 640, and serial 21640. This one is thought to have served with the Royal Canadian Navy. It was bought by OH-B and registered as CF-EHB, and arrived at Southend shortly after CF-IHB. It was registered as G-WGHB, the initials of Hayden's brother, and later moved to Duxford.

T-33C N12420 This aircraft's background is unknown, it arrived from Italy on March 9th 1976 and has yet to be registered in the United Kingdom. It is Lockheed built.

Manuel Willow Wren BGA.162

Built 1932, single seat glider.

This was the prototype Willow Wren, first flown in December 1932. It was developed from the Crested Wren of 1931. W. L. Manuel built two Willow Wrens and a further five were built from plans supplied by the designer. The Wrens were painted in different colours, and were generally known individually as the 'White Wren', 'Blue Wren' etc. BGA.162 was therefore also known as the 'Yellow Wren'. It was owned by the designer for some time after its Certificate of Airworthiness was issued in September 1934, but was later sold to the Cambridge Gliding Club and then to L. A. Read. After a period in storage it was bought by N. H. Jones of the Tiger Club and restored by W. L. Manuel at Rochester in 1964 - however it later returned to storage, until bought by Chris Wills and Mike Russell in December 1974 for rebuild to flying condition. It was moved to Duxford in late 1976, and is now owned only by Mike Russell.

Messerschmitt Me 163B-1 Komet 191060

Built 1942/3, single seat interceptor, powered by one Walter 109.509 bi-fuel rocket motor.

This aircraft was captured in Germany in 1945 and subsequently brought to the Royal Aircraft Establishment at Farnborough. Here it wore RAF markings, including the yellow ringed 'P' prototype markings and the serial VF241. It was flown extensively from Farnborough towed behind a Spitfire for aerodynamic investigation in 1945 and 1946. It was struck off charge at Wittering in December 1947, but was taken to No.54 Maintenance Unit at Cambridge for storage, and was later transferred to the museum at RAF Cranwell. When this museum closed the aircraft was loaned to the Imperial War Museum and it was displayed at Lambeth in near original paint scheme. It was painted up as Werke Nr. 191160 - this is thought to be an error. It was moved to Duxford by road in October 1976.

Mignet HM.14 Pou du Ciel

Built 1930s, single seat ultra light, powerplant unknown.

This aircraft is one of perhaps over a thousand started in the United Kingdom in the 1930s. This particular aircraft has so far yielded up none of its past. It is owned by the Essex Aviation Group and is kept in their workshop on the airfield. A completely new set of wings have been built for the aircraft, using the wings of the Shuttleworth Trust's example (G-AEBB) as a pattern. Any details welcomed on this machine's history.

Nord N.1101 Noralpha G-ATIX

Built late 1940s, constructor's number 154, four seat cabin monoplane, powered by one Renault4Q 10 in-line piston engine of 233 horse power.

This aircraft was built by Nord at Les Mureaux for the French Armed Forces, before joining the register as F-BLYO. It was imported into the United Kingdom in February 1966 and registered as G-ATIX. It has been with several owners since, the current owners being J. Reast and B. Cope. The N.1101 is a production version of the Messerschmitt Me 208 and the current owners have painted up G-ATIX to represent a Luftwaffe version. It is airworthy.

North American Harvard IIB G-BBHK

Built 1942, constructor's number 14A-787, advanced trainer, powered by one Pratt and Whitney R-1340-49 radial piston engine of 600 horse power.

This aircraft was built in Canada by the Noorduyn Aircraft Company, originally to United States Army Air Force order, serial number 42-12540, but was quickly diverted to the RAF for use in the Empire Training Scheme. With the RAF it was given the serial number FH153. Post-war it was sold to the Netherlands Air Force and it became B-158. It was disposed of in the early 1970s and was stored for a while. It was bought on behalf of Personal Plane Services and registered in Holland for delivery to Booker. At first the registration PH-HTC had been officially allotted but this was not worn and the more fitting PH-PPS was used. PPS brought the aircraft up to full civil standards and it was registered as G-BBHK for its current owner, Rob Lamplough. It was moved to Duxford in the first week of January 1977 and is painted in post-war training colours.

North American P-51D Mustangs 41, 43 and 146

Built 1940s, single seat fighters, powered by one Packard Merlin V-1670-7 of 1,695 horse power each.

These three aircraft are the property of Rob Lamplough and have been obtained from various sources in Israel, where they last served operationally. One aircraft, 43, is currently undergoing a full rebuild to airworthy condition. The other two will hopefully follow. Number 43 arrived at Duxford from Booker (where it had been briefly) on July 18th 1976 along with the wings of the Yakolev 11 (see below). The other two arriving in late 1976. Little is known about these aircraft, except that at least one was involved in the Suez Campaign, as it is bedecked in very crude identification stripes from that incident. All are believed to be from a batch of P-51s supplied from the Swedish Air Force in 1951, the Mustang being known as the J26 in Swedish service.

North American P-51D Mustang '472258' 'IW-Z'

Built 1945, single seat fighter, powerplant as above.

Duxford's famous 'Big Beautiful Doll' is painted to represent the mount of Colonel John B. Landers, Commanding Officer of the 78th Fighter Group, United States Army Air Force, based at Duxford on bomber escort duties in 1945. It was really built for the USAAF as 44-73979 at the Inglewood plant of North American and was delivered for service on April 26th 1945. It served at Kelly and Hobbs Air Force Bases in the U.S.A. In 1947 it was sold to the Royal Canadian Air Force as 9246, and was at one time coded '612'. It served in the RCAF with No.402 'Winnipeg' Squadron and No.420 'London' Squadron. It was struck off official charge in May 1951 after an accident. It was then put on display on the gate of St. John Air Force Base, in Quebec until being presented to the Imperial War Museum as a gift from the People of Canada. It moved to Duxford in December 1971 and commenced its refurbishment to the immaculate condition it is displayed in today.

North American B-25J Mitchell N7614C

Built 1944, aerial camera platform, powered by two Wright R-2600-29 radial piston engines of 1,700 horse power each.

This aircraft started life as a B-25J medium bomber for the United States Army Air Force, serial number 44-31171. Its service history is unknown and its next certain happening was its open storage at Luton airport in 1970 after use in several aerial filming contracts in the United Kingdom and Eire. In this guise, as N7614C, it was under the ownership of Jeff Hawke. Since then it has earned itself a considerable reputation as a craft of ill-repute. On an abortive ferry flight to the U.S.A. it was impounded at Prestwick Airport and was later similarly treated at Dublin airport, in both cases gun-smuggling was mooted at. It flew into Shoreham airport in 1974, amidst great speculation that it was to be completely refurbished. This never happened and the aircraft languished at Shoreham and only a very patient view by the Airport prevented this valuable aircraft from being reduced to ingots. The aircraft was rescued in 1976 by the Imperial War Museum, a DAS team taking six weeks to dismantle and remove it to Duxford. It is to be restored to World War Two colours and an original nose position has already been obtained.

North American F-100D-11-NA Super Sabre 42165 '11-ML'

Built 1954, single seat fighter, powered by one Pratt & Whitney J57 turbojet of 11,700 lbs st.

This aircraft was built by the North American Inglewood plant for the United States Air Force with the serial number 54-2165 as an F-100D-10-NA, later receiving systems modifications to render it an -11-NA. It was supplied to the French Air Force after a brief career with the USAF. It served the French using its USAF serial. It was first in use with Escadre 1/3, with whom it was coded '3-IG' and used the call-sign 'FUGIG'. It was then with Escadre 2/11 and coded '2/11', call-sign 'FUGMB'. The aircraft is displayed today in these unit markings. In 1975 it was retired from service, and being supplied under American aid, had to be returned to the USA. The USAF had no further use for the aircraft and it was decided to fly it to Sculthorpe for disposal, along with many other examples. It flew into Sculthorpe on November 24th 1975 using the call-sign 'FSDML' and the code '11-ML'. This being the call-sign of the Escadre de Convoyage '070' based at Chateaudun, who were responsible for ferrying the 'Sleds' to Sculthorpe. It was stored at Sculthorpe until May 17th 1976 when it was taken by road to Duxford to join the Imperial War Museum's growing collection.

Percival Q.6 Petrel G-AFFD

Built 1938, constructor's number Q.21, general purpose twin, powered by two De Havilland Gipsy Six II in-line piston engines of 205 horse power originally.

G-AFFD was the second Q.6 to be built, and was built at Luton to the order of Sir Philip Sassoon, who based it at Lympne. It was impressed on April 4th 1940 and served with No.6 Anti-Aircraft Co-operation Unit until November 25th when it was flown to No.39 Maintenance Unit at Colerne and serialled X9407. From Colerne it saw service at Northolt, Heston, Halton, Old Sarum, No.510 Squadron at Hendon, and the Metropolitan Communication Squadron until April 11th 1946 when it was taken to No.5 MU at Kemble for disposal. It was bought by the Yorkshire Aero Club at Sherburn-in-Elmet who registered it as G-AIEY until they discovered its original identity. It was withdrawn from use at Redhill in August 1956 and became the property of the College of Aeronautical Engineering there and used as an instructional airframe for many years. During this time it lost its outer wings. It was acquired by the Midland Aircraft Preservation Society in 1974 and moved to Duxford for storage.

Percival Proctor I R7524

Built 1941, light cabin monoplane, one De Havilland Gipsy Queen II of 210 horse power.

This aircraft was built for the Royal Air Force as R7524 by F. Hills and Sons at Trafford Park, Manchester. It was demobbed in 1947 and registered as G-AIWA. It has had a string of private owners since, and is currently in the hands of T. Storey and M. Baraclough who have restored it to its immaculate silver hue and have based it, fully airworthy, at Duxford since 1975.

Royal Aircraft Factory RE.8 F3556

Built 1918, two seat artillery observation aircraft, powered by one RAF 4A piston engine of 150 horse power.

This aircraft was built by the Daimler Motor Company and was funded by the People of Ceylon and christened 'A Paddy Bird from Ceylon' at their request. It was shipped to France and assembled ready for action, but was to achieve only thirty minutes flying time in France on October 31st 1918. It was put back into its crate and stored. It was presented to the Imperial War Museum (still in its crate) and was displayed at Lambeth until moving to Duxford in 1975 for complete restoration to static condition. This long job is continuing.

Short S.25 Sunderland MR. Mk.5 ML796 '50-s-3' see photo 14

Built 1944, long range maritime patrol flying boat, powered by four Pratt and Whitney R-1830-90B Twin Wasp piston engines of 1,200 horse power each.

Ml796 was built at Rochester and delivered to Calshot on May 15th 1945, entering service with No.228 Squadron at Pembroke Dock soon after. It later moved on to No.4 Operational Training Unit from March 3rd 1946. It was retired and flown to No.57 Maintenance Unit (detachment) at Wig Bay on July 10th 1946 but was re-issued to service flying supplies into blockaded Berlin from Finkenwerder during the Berlin Airlift of 1948/49. It was on charge with No.230 Squadron for this operation. On December 20th 1949 it was earmarked for sale to the French Navy and was handed over to Short Brothers and Harland at Sydenham on June 22nd 1950 for conversion and overhaul, this being completed on July 31st 1951. It was delivered to No.57 MU on August 3rd 1951 and was delivered shortly afterwards to the Aeronavale. It saw service with 7FE and later 27F (the markings of which it still wears) at Dakar until being withdrawn from use in 1960. It was brought to France and stored at Lanveoc-Poulmic. Its flying days were now over, the big 'boat having clocked a total of 2,400 hours in the air. It was bought by Mr. R. Bertin in 1965 and moved to Maisden-le-Riviere and used as a disco. In 1969 it was moved to La Baule and continued its new role of night-club and restaurant. Costs of keeping the concern going and the imminent arrival of a motorway through the site of the Sunderland led Mr. Bertin to offer it free to anyone who could remove it. After an inspection a team from the DAS was despatched to start work on June 24th 1976 and the hard task of bringing the craft back to Duxford and the Imperial War Museum began. ML796 was moved on five trucks and took the Dunkirk-Harwich ferry, finally arriving at Duxford on July 9th. Its restoration will take many years as the only way to move the huge hull was to cut it horizontally down the centre-line.

Slingsby T.13 Petrel 1 BGA.651

Built 1938, constructor's number 361, single seat glider.

This aircraft was the last of three Petrels to be completed and was registered with the British Gliding Association as certificate number 651 in 1950, although it was at one time G-ALPP. In 1954 it was sold in Eire as IGA.101, and later became EI-101. It was retired in 1969 and was stored at Dublin Airport for the Irish Aviation Museum. It was returned to the United Kingdom in November 1973 and flown from Duxford by Mike Russell. It is now part of the collection of gliders owned by Mr. Russell, the Petrel being co-owned with Professor J. V. Byrne.

Supermarine Spitfire Vb AR501 'DN-N'

Built 1942, single seat fighter, one Rolls-Royce Merlin 45M of 1,440 horse power.

Built by Westland Aircraft, AR501 was delivered to No.8 Maintenance Unit at Little Rissington on June 22nd 1942, moving to No.6 MU at Brize Norton on July 7th. It was delivered to No.310 Squadron at Exeter on July 19th and became the mount of Squadron Leader F. Dolezal D.F.C. whose markings it currently wears. It suffered Category B (Beyond repair on site) damage and was issued to No.67 MU Taunton for repair on March 22nd 1943. This was too severe for the MU as it was handed back to the makers the following day. It was ready for action again on July 3rd, joining No.33 MU at Lyneham two weeks later. On August 16th it was issued to No.3501 Support Unit and two days later to No.504 Squadron at Church Fenton for a mere two days before going on charge with the Church Fenton Station Flight. On October 16th 1943 it re-entered squadron service with No.312 at Ibsley, followed by No.422 Squadron from February 27th 1944, No.58 Operational Training Unit from March 30th 1944, No.1 Tactical Exercise Unit from April 26th 1944 and No.61 OTU from July 4th 1944. On September 9th AR501 received another Category B blow and was entrusted to Air Service Training Limited for repair from September 22nd until November 23rd. It was back at No.33 MU on December 2nd. It was not issued for service until April 24th 1945 when it joined the Central Gunnery School for its last operational flying, being taken to No.29 MU at High Ercall on August 26th 1945. It was acquired by Loughborough College on March 21st 1946 and installed as an instructional airframe there. It was acquired by the Shuttleworth Trust in June 1961 and moved to Old Warden. In 1967, like many other Spitfires, it was readied to become a star in the film 'The Battle of Britain'. For this purpose it was registered as G-AWII on April 25th 1968. During the film (some of the shooting of which was done at Duxford) it appeared as 'N3314', N3316','N3320' and was coded 'AI-E', 'AI-G' and 'DO-A'. It was moved to Duxford on May 7th 1973 and underwent a painstaking rebuild to full flying trim. It first flew in its new colours on June 27th 1975 and is based at Duxford by the Shuttleworth Trust.

Supermarine Spitfire LF. Mk. IXe

Built 1944/45, single seat fighter, one Rolls-Royce Merlin 76 of 1,720 horse power.

This aircraft is owned by Bob Lamplough, and, like his Mustangs, is from Israel, having been flown operationally with the Israeli Air Force. Its history is not known, although it is most likely that it was acquired in 1946 from the Czech Air Force and is ultimately ex RAF. It arrived at Duxford in late 1976 and is under restoration to, hopefully, flying trim. The serial TE566 has been quoted for this aircraft.

Supermarine Spitfire PR.XI PL983

see photo 11

Built 1944, single seat photographic reconnaissance aircraft, one Rolls-Royce Merlin 70.

First delivered to No.6 Maintenance Unit at Brize Norton on October 31st 1944, PL983 was to have a very varied flying life, including being the personal mount of the United States Air Attache and being on the American civil register. It saw service with, No.1 Pilot's Pool (PR HQ) at Benson starting November 24th 1944, No.34 Wing from January 5th 1945, No.4 Squadron from February 22nd, No.2 Squadron from September 6th. It was back at No.6 MU on January 12th 1946, then was shuttled to No.39 MU Colerne on July 15th, back to No.6 MU on July 29th and finally to No.33 MU at Lyneham. From July 22nd 1947 it was placed on Contract Loan to Vickers Armstrong at Eastleigh for loan to the United States Air Attache. On January 27th 1948 it was officially handed over to Livingstone Satterthwaite, the American Civil Air Attache, by Vickers at Eastleigh. It was registered N74138. It was flown from Eastleigh to join the U.S. Embassy Flight at Hendon that day by Miss Lettice Curtis. On August 28th Lettice was to fly the Spitfire, race number 4, in the Lympne Air Races. In this she was placed 5th, pipped to the post by a two-seat Spitfire, two Vampires and a Firefly. That day Lettice also broke the 100 kilometre closed circuit record with a speed of 313·208 m.p.h. It was returned to Vickers at the end of Mr. Satterthwaite's office, and was ear-marked for the Shuttleworth Trust, to whom it was ferried in 1950 in the class B markings G-15-109. It was repainted in PR colours at Old Warden and was displayed statically for many years. It was moved to Duxford by road on August 30th 1975 and is well into a rebuild to flying condition that will take many years of dedicated skill.

Vickers 668 Varsity T. Mk.1 WF425

Built 1952, constructor's number 568, meteorological research aircraft, powered by two Bristol Hercules 264 radial piston engines of 1,975 horse power each.

Built at Hurn, WF425 flew for the first time on August 4th 1952. It was delivered eight days later to No.2 Air Navigation School, becoming 'F'. This was followed by service with No.1 ANS, this time as 'M', and with the Central Flying School at Little Rissington, as '18'. It was retired to No.5 Maintenance Unit at Kemble in September 1969. It was transferred to the Royal Aircraft Establishment and delivered to the Meteorological Research Flight at Farnborough on March 24th 1970. It served in this role until replaced by a modified Hercules, and was withdrawn from use in 1974. It was acquired by the Imperial War Museum and flown into Duxford for static display.

Vickers 668 Varsity T. Mk.1 WJ945/G-BEDV

Built 1953, multi-purpose trainer, powered as above.

WJ945 was also built at Hurn, and flew first on February 23rd 1953. It was delivered to 'A' Flight of No.116 Squadron on March 11th 1953, this unit being renumbered No.115 Squadron on August 21st 1958. Like WF425 above, it also served with the Central Flying School at Little Rissington, coded '21'. It was acquired by the DAS Chairman, D. Selway and flown into Duxford in October 1974. It was then refurbished to become the flag-ship of the DAS, and flew its test flight for its Permit to Fly on July 26th 1976 in time for it to appear at the Greenham Common display that year. It is intended to visit many air displays with this aircraft in the coming years. It has been allocated the civil registration G-BEDV, and this is carried in minute letters on the tail section.

Vickers 701 Viscount G-ALWF

see photo 19

Built 1952, constructor's number 5, medium range airliner, powered by four Rolls-Royce Dart R.Da.3 Mk. 506 turboprops of 1,547 estimated shaft horse power each.

G-ALWF was registered to Vickers-Armstrongs on January 2nd 1950 but did not fly, from Weybridge, until December 3rd 1952. After acceptance trails, it was delivered to British European Airways Corporation at Heathrow on February 2nd 1953, being named 'Sir John Franklin'. It served steadily with BEAC until phased out of service (replaced by Viscount 800s) in 1964. It was sold to Channel Airways in March and was based at Southend. From November 18th 1964 it was leased to British Eagle International Airlines, and named 'City of Exeter', being returned to Channel in May 1965. It was delivered to Cambrian Airways at Rhoose on December 15th 1965 and was to serve with Cambrian until its announced retirement in 1972. At this stage it was realised that 'Whisky Fox' was the oldest surviving Viscount in the World, and that it was therefore most worthy of preservation. To this end the Viscount Preservation Trust was formed and the Corporation of Liverpool kindly gave free hangarage space so that the aircraft could be properly preserved and at the same time displayed to the public. It was flown on a special publicity flight from Rhoose to Heathrow to Speke on April 12th 1972. This was its last flight and the aircraft was officially withdrawn from use that day. G-ALWF had completed 28,299 hours in the air, from some 25,938 flights. G-ALWF was opened by the Mayor of Liverpool to public display on December 5th 1972 and it was to become a popular attraction until new security regulations at Speke brought a stop to public access, and to the funds needed to keep the project alive. G-ALWF's future looked bad, although it managed to earn its keep for a while by donating its outer wings to one of the Series 700s on the Prestwick feeder run with BOAC. A pair of time-expired wings were taken in exchange. In the Autumn of 1975 the DAS took an interest in the airliner and the rescue was 'on'. From October 11th 1975 until February 15th 1976 a team from the DAS worked on 'WF most weekends in the long task of dismantling it. The fuselage was moved to Duxford by road on February 21st 1976 with the wings and other trappings following shortly. The work of restoring G-ALWF is continuing.

Westland WS-55 Whirlwind HAS. Mk.7 XG577

Built 1957, anti-submarine naval helicopter, powered by one Alvis Leonides Major 755.

XG577 first took flight from Yeovil on March 28th 1956 and was delivered to Royal Naval Air Station Lee-on-Solent on April 16th. From August 8th it was on charge with the Ship's

22 *Corsair KD431 of the Fleet Air Arm Museum, flaps down, at Yeovilton* *via Les Hunt*

23 *Early production Buccaneers XK489 & XK488 (with FAAM) in 1960* *•'Flight' 40811 S*

24 *Concorde 002 G-BSST makes its maidên flight from Filton 9/4/69* *'Flight' D9-29/12*

25 *Duxford's Shackleton XF708 in the colours of No.120 Squadron 'A'.* *S. G. Jones*

26 *Beverley XB261 en route Boscombe Down-Rochford October 6th 1971.* *'Flight' 71-6353*

27 *The HAM's Dragon G-ACIT near Old Warden in August 1971.* *'Flight' L71-1493/27a*

Flight of the aircraft carrier H.M.S. 'Albion', with the code '979'. This was short-lived as it was taken on charge with No.701 Squadron 'C' Flight, but still on the 'Albion', from October 28th. By November 13th it was with 'B' Flight on temporary loan, embarked upon H.M.S. 'Ark Royal'. Twelve days later it was at the Royal Naval Aircraft Yard at Donibristle. It was issued to the helicopter training unit, No.705 Squadron, at Lee-on-Solent on July 24th 1958 and wore the code '531' followed by '523' from September 1958. Its career then became one of shuttling between units, fulfilling different roles :- No.815 Squadron as '301' from January 6th 1959 initially at Eglinton, then at Portland, then it is thought to have been transferred to No.737 Squadron (which was renumbered from No.815) as it was recoded '762', in December 1961 it re-joined No.705 Squadron, now at Culdrose, and was coded '528'. On March 9th 1967 after storeage at Culdrose from 1964, it was allocated to Arbroath for ground instructional duties, as A2571/E4582. It was struck off charge on September 27th 1967 but moved to Lee-on-Solent by 1970. It was acquired by the Imperial War Museum and moved there by road on June 13th 1974. It was damaged in unloading by the RN team in charge, and a replacement was arranged as detailed below. XG577 remains at Duxford however.

Westland WS-55 Whirlwind HAS. Mk.7 XK936

Built 1957, role and powerplant as for XG577.

This aircraft was transferred to Duxford on November 25th 1975 from Royal Naval Aircraft Yard Wroughton as a replacement for the damaged XG577, see above. It first flew on July 18th 1957 at Yeovil and was first delivered to Lee-on-Solent on August 1st. On the 23rd it joined No.845 Squadron, and was apparently not given a code. On December 27th it was moved to Eglinton and No.820 Squadron, still apparently uncoded. By May 1958 it was with No.701 Squadron and served as such until April 27th 1960 when it was transferred to No.848 Squadron on H.M.S. 'Bulwark' for a tour to the Far East. It was stored at Tengah from Janaury 4th 1961 until going back onto the 'Bulwark' on December 29th. It returned to the United Kingdom on December 14th 1962, at Culdrose. It joined No.847 Squadron at Culdrose on October 14th 1964, still without apparently having worn a code. This was remedied in October 1966 when it joined No.705 Squadron at Culdrose where it gained the markings '554'. By 1969 these had been changed to '562'. It was removed to Wroughton on December 17th 1974 and was stored until allocated to the Imperial War Museum.

Notes : The following aircraft are not complete, or are considered to be transitory, and are listed only in brief format :-

Auster J/1N Alpha G-AIBR of the DAS. c/no. 2151, registered in 1946. Damaged beyond repair at Gamston in September 1970, stored at Sywell. Moved to Duxford in late 1975.
De Havilland Canada DHC-1 Chipmunk T.10 8341M with Ormond Haydon-Baillie, for export. From No.1 School of Technical Training at Halton in late 1976, ex WP963 and No.229 OCU '1'.
Gloster Meteor F.4 VT229 with Jeff Hawke, thought to be for export to the U.S.A. Purchased from the Colerne Collection sale in early 1976, maintenance serial 7151M.
Gloster Javelin FAW.9 XH892 with Jeff Hawke, thought to be for export to the U.S.A. Bought along with VT229 from Colerne, maintenance number 7982M.
Heinkel He 111 of the IWM. Many components salvaged from a diving expedition in Norway.
Martin B-26C Marauder 41-35253 of the IWM. Forward fuselage and other components salvaged in 1975 from a scrap-yard in Warrington.
Messerschmitt Bf 109G 14120 Discovered in Norway with the He 111. Forward fuselage currently under restoration.
Messerschmitt Bf 110 rear fuselage of Rudolf Hess's aircraft, coded 'VJ-OQ', in which he flew to Scotland from Augsburg to Scotland on May 10th 1941 on a peace mission. With the IWM.
Yakolev Yak-11 ٥٩٠ (590) with Bob Lamplough. Bought in Israel and shipped to U.K. in 1976. Wings only at Duxford. Fuselage at Booker on overhaul. Will eventually fly from Duxford. Served with the Egyptian Air Force.

Ormond Haydon-Baillie currently has Gloster Meteor F. Mk.8 WH291 in store at No.5 MU at Kemble, this aircraft will join his fleet at Duxford in due course.

FLEET AIR ARM MUSEUM

Since its formal opening in 1964 by His Royal Highness The Duke of Edinburgh, its present patron, the Fleet Air Arm Museum's collection has increased to an extent whereby a large scale development programme has become necessary. A public appeal launched in 1973 allowed Phase One, a new exhibition hall, to be completed by October 1975, and a year later work began on Phase Two, the workshops and office complex.

Attendance figures in 1976 were about a quarter of a million, a large advance on other years, due in a large measure no doubt to the arrival of the Science Museum's Concorde 002.

The Museum (on the B3151 two miles east of Ilchester) is open all year except for two weeks over Christmas and New Year. The hours are 1000 to 1730 on weekdays (including Bank Holidays) and 1230 to 1730 hours on Sundays. During the winter months closing is at sunset when this is before 1730 hours. There is late opening to 2000 on Thursdays from May to September inclusive. Admission charges are modest and there are special rates for school coaches. There is an additional charge to view the interior of Concorde.

There are research facilities for the serious student and application should be made in advance to ensure that material is readily available. Copies of photographs and documents can be supplied at moderate costs.

The nearest railway stations are Sherborne, Yeovil Junction and Yeovil Pen Mill. Bus service No.478 from Yeovil passes the gate.

Address : The Fleet Air Arm Museum, Royal Naval Air Station, Yeovilton, Ilchester, Somerset BA22 8HT. (Correspondence addressed to the Director.)

Telephone : 093584 551 Extension 528 or 531.

Compilers' note : Three incomplete airframes are displayed inside the Museum, these are dealt with in briefer format at the end, along with aircraft held in reserve.

BAC/Aerospatiale Concorde 002 G-BSST see photo 24

Built 1969, constructor's number 13520, production number 002, prototype super-sonic airliner powered by four Rolls-Royce/SNECMA Olympus 593 turbojets of 35,080 pounds static thrust.

This aircraft, owned by the Science Museum, flew into Yeovilton on March 4th 1976 at the end of some 800 hours of test flying, preparing the way for production Concordes to enter service. Of that time some 200 hours were spent at super-sonic speeds. It is displayed on a special ramp and it was officially opened to the public on July 26th 1976. G-BSST was the second Concorde to fly, the first to be assembled in the United Kingdom. It flew for the first time on April 9th 1969 piloted by Brian Trubshaw, from the production centre at Filton to the flight test centre at Fairford. It has been involved principally in engine and basic aerodynamic trials, and was never intended to be sold for airline use, being unrepresentative of the production versions.

Blackburn NA-39 Buccaneer S. Mk.1 XK488 see photo 23

Built 1958, prototype naval strike aircraft, powered by two Bristol Siddeley Gyron Junior 101 turbojets of 7,100 pounds static thrust each.

This aircraft was part of a batch of twenty pre-production aircraft and was to spend its life in test and development work. Indeed it has the dubious honour of being the only Buccaneer S. Mk.1 never to wear the wording 'Royal Navy'. It first flew from Brough on October 31st 1958 and was initially used by the manufacturers for engine development work. It moved to Hatfield for similar work with De Havillands for the period November 1960 to April 1961. From July 1962 it was with Bristol Siddeley Engines at Filton until at least April 1964. By May 1968 it was on the gate at RNAS Yeovilton and was soon brought into the Museum's fold.

Blackburn NA-39 Buccaneer S. Mk.1 XN967

Built 1963, two seat naval strike aircraft, powerplant as above.

This aircraft was delivered from Brough to Holme-on-Spalding-Moor on March 29th 1963 and after flight testing was delivered to the Aircraft Holding Unit at Lossiemouth on April 24th 1963. It became operational with No.809 Squadron in August becoming '234/LM', by mid-1964 it was recoded as '228/LM'. On March 26th 1965 No.809 Squadron was reduced to second-line status and renumbered No.736 Squadron. XN967 was still with the unit and was recoded '640/LM' accordingly. It later became '642/LM' and then '630/LM' - the latter from October 7th 1968. In December 1970 the type was grounded and XN967 was transferred to the charge of the Naval Aircraft Servicing Unit at Lossiemouth. It left Lossie on November 19th 1971 for Yeovilton and the Museum.

Chance-Vought Corsair IV KD431 'E2-M'

see photo 22

Built 1944, constructor's number 1871, single seat naval fighter, powered by one Pratt and Whitney R-2800 radial piston engine of 2,100 horse power.

This aircraft was built to United States Navy standard as a FG-1A by Goodyear at their Akron, Ohio, plant. It was diverted on the line to the Royal Navy. It arrived at the Aircraft Holding Unit at Coimbatore in late 1944. It is thought to have been issued first to No.731 Squadron at Royal Naval Air Station East Haven, coded 'E2-M' from November 1945. It later served with No.716 Squadron until being retired to Lossiemouth for storage. It was bought by the College of Aeronautics at Cranfield and used by them for instructional purposes from November 1946. It was bought by the now defunct Historic Aircraft Preservation Society in 1963 and refurbished for display at Yeovilton from 1964. The aircraft appears in its 1945 service markings and is the property of the Museum.

De Havilland DH.82A Tiger Moth XL717

Built 1940, constructor's number 83805, basic trainer, powered by one De Havilland Gipsy Major in-line piston engine of 130 horse power.

XL717 has had a varied history, and has only been with the Navy since 1956, despite its age. It was built by Morris Motors at Cowley for the Royal Air Force as T7291 and was officially taken on charge with No.33 Maintenance Unit at Lyneham on August 29th 1940. It was first issued to No.19 Elementary Flying Training School on April 4th 1941, followed by No.24 EFTS at Rochester, coded 'FJGR' from February 7th 1942 until it returned to No.33 MU on July 30th 1946. It was sold off on September 3rd 1953 to Transair Limited, but not put on the civil register, as G-AOXG, until October 1956 with Hants and Sussex Aviation. Life as a civilian was very brief as it was with H&S to fulfil a contract for the Fleet Air Arm who were short of Tiger Moths to serve with Station Flights for glider towing duties. On November 7th T7291 re-entered military service, this time with the Navy and a new identity - XL717. It was first used by the Royal Naval College, Dartmouth, and then from July 1966 with the Station Flight at Lossiemouth where it put in much time with the Fulmar Gliding Club from nearby Milltown. In 1972 it joined the Museum.

De Havilland DH.100 Sea Vampire I LZ551/G

Built 1945, prototype development aircraft, one De Havilland Goblin turbojet.

This aircraft has particular significance to the Fleet Air Arm. It was the first 'hooked' jet aircraft and carried out the first ever landing of a jet powered aircraft on an aircraft carrier. Piloted by Lieutenant Commander E. M. Brown LZ551/G landed on H.M.S. 'Ocean' on December 3rd 1945. This aircraft was built to Specification E.6/41 by the English Electric Company. The serial carried the suffix /G to indicate that it was to be guarded at all times whilst on the ground. LZ551/G soon became the property of the Science Museum and it was held in store before going on display at Yeovilton.

De Havilland DH.110 Sea Vixen FAW. Mk.1 XJ481

Built 1958, two seat test and calibration aircraft, powered by two Rolls-Royce Avon 208 turbojets of 10,000 pounds static thrust.

XJ481 first flew on September 2nd 1958 and was to enter a long career of test work. It was delivered to the Handling Squadron of the Aeroplane and Armament Experimental Establishment at Boscombe Down on November 28th 1958. It was detached on H.M.S. 'Centaur' for trials during 1959. It was shipped to Australia on March 13th 1960 for use with the Long Range Weapons Establishment at Woomera on weapons trials. It stayed in the desert until March 1967 when it came home by ship. It arrived by June and joined Hawker Siddeley Dynamics. It was back with the A&AEE from November 6th 1968 and served with 'C' Squadron on missile trials, ending with the Martel. During this time it was painted in the black and white calibration markings that it wears today. It was retired and despatched to Yeovilton in March 1974 to take its place in the Museum.

De Havilland DH.110 Sea Vixen FAW. Mk.2 XS590 '131/E'

Built 1966, two seat all weather naval fighter, powerplant as above.

This aircraft was delivered direct from Hawarden to No.892 Squadron at Yeovilton on March 9th 1966. It was coded '303/H' (No.892 operating on H.M.S. 'Hermes') and often formed part of the Squadron's aerobatic team 'Simon's Sircus'. It moved to Royal Naval Aircraft Yard Sydenham in February 1968 and on to the Naval Aircraft Servicing Unit at Brawdy in early 1969. It joined No.899 Squadron on March 12th 1969, taking up the markings it wears today.

It stayed only for a short term with No.899 Squadron, going to the NASU at Yeovilton in May 1970. It was officially presented to the Museum on November 26th 1970, although it was some time before it actually was towed over to the Museum area for display.

De Havilland DH.112 Sea Venom FAW. Mk22 WW138 '227/0'

Built 1955. two seat all weather fighter, powered by one De Havilland Ghost of 4,850 lbs s.t.

Built by Airspeed at Christchurch, WW138 was taken on charge at Stretton on March 2nd 1955. It was first in service with No.809 Squadron at Yeovilton from September 6th 1955, with the coding '227/O' - the markings it wears today in the Museum. It was on tour in the Mediterranean from November 15th 1955 until March 1956. It was then taken to the Royal Naval Aircraft Yard at Sydenham in July, before going back to the makers for an unspecified reason in January 1957. It was held at Stretton from July 1957 until January 1958. It moved to Abbotsinch for modifications from July 1958 until January 1959. It was back at Sydenham by July 1959 and was officially converted to FAW. Mk.22 standards (having been an FAW. Mk.21 until then) by November 1st 1959. It was in store at Lee-on-Solent from January 1960 until the following January at least. It was at Culdrose in July 1961 and WW138 toured the Mediterranean with No.809 Squadron again. From April 4th 1962 it joined No.831 Squadron at Watton and Culdrose on electronic counter measures work, coded '383' by September 1963. It went into store at Sydenham from January 1965 for a year. In July 1968 it joined the Air Weapons School at Yeovilton, becoming '734/VL' by April. It was retired in September 1969 and joined the Museum in 1969.

De Havilland DH.115 Sea Vampire T. Mk.22 XA129

Built 1954, advanced trainer, powered by one De Havilland Goblin 35 of 3,500 lbs. s. t.

XA129 started its service life with No.736 Squadron based at Lossiemouth from August or September 1955, coded '240/LM'. It later went on to be the mount of the Flag Officer Flying Training, gaining the code '945'. By July 1968 it joined the Air Director School at Yeovilton and was used as a continuation trainer. It was retired by 1970 and presented to the Museum.

Douglas Skyraider AEW. Mk.1 WT121 '415/CU'

Built 1951, airborne early warning aircraft, powered by one Wright Double Cyclone R-3350-26WA radial piston engine of 3,300 horse power.

WT121 was built as an AD-4W for the United States Navy, serial number 124121 and first flew on January 24th 1951. It first served with VC-11 until August 1952, then with VC-12 from October 1952 followed by service at Naval Air Station Quonset Point. It was flown to NAS Norfolk on March 9th 1953 for re-allocation to the Fleet Air Arm. It was shipped to the United Kingdom on September 15th 1953, arriving at Abbotsinch on the 29th. It was delivered to Culdrose on March 18th 1954 and was issued to No.849 Squadron 'E' Flight, coded '326'. 'E' Flight later became 'D' Flight, with WT121 still on charge. It was flown to Hal Far on August 17th 1955 and held there in readiness until being flown back to the United Kingdom on June 22nd 1956, arriving at Culdrose again on July 6th. It rejoined 'D' Flight as '428', later gaining the tail code 'B'. It was loaned briefly to 'C' Flight in November 1957. On October 30th 1958 it was issued to the Headquarters Flight of the Squadron and was coded '415'. It was retired on September 30th 1960 and placed in the Aircraft Holding Unit at Culdrose. It was refurbished & painted in its last operational colours ready for display at the Fleet Air Arm Museum. It was flown underneath Sea King HAS. Mk.1 XV698:592/CU on April 21st 1972 from Culdrose to the Royal Fleet Auxiliary 'Engadine' and then to the south coast and again by Sea King to Yeovilton.

Fairey Swordfish II 'W5984'

Built 1943, three seat torpedo strike aircraft, powered by one Bristol Pegasus 30 radial piston engine of 750 horse power.

This aircraft is displayed in the markings of the aircraft flown by Lieutenant Commander Esmonde V.C. D.S.O. in the attack on the 'Scharnhorst' and the 'Gneisenau' in 1942. Underneath this paint scheme lies a Blackburn built Swordfish (often called a 'Blackfish') HS618 which was delivered from Sherburn-in-Elmet in April 1943. Little is known of its history except that it was with the Ferry Pool at Donibristle from May 1943. It was kept at the Royal Naval Engineering College at Manadon, where it was allocated the maintenance airframe number A2001. It joined the Museum at an early stage.

Fairey Fulmar II N1854

see photos 21 & 52

Built 1939, constructor's number F.3707, two seat naval fighter prototype, powered by one Rolls Royce Merlin 30f of 1,300 horse power.

This aircraft was built at the Heaton Chapel plant of Fairey's and was flown for the first time from Ringway on January 4th 1940 powered by a Rolls-Royce Merlin VIII of 1,080 horse power. It was used throughout the war for manufacturer's trials, followed by trials at the Aeroplane and Armament Experimental Establishment at Boscombe Down. The latter included deck landing trials on H.M.S. 'Illustrious'. It returned to Faireys in 1945 and was civilianised at Ringway in 1946 with the engine noted in the heading. It was given the registration G-AIBE and used by the makers as a 'hack'. In June 1959 it was repainted from its civil scheme into more representative camouflage and military markings. It made its last flight on December 18th 1962. It was soon donated to the Navy and kept at Lossiemouth until joining the Museum on September 22nd 1972.

Fairey Firefly TT. Mk.4 VH127

see photo 20

Built 1947, constructor's number F.8026, target tug, powered by one Rolls-Royce Griffon XII piston engine of 1,815 horse power.

VH127 was built as an FR. Mk.4 fighter reconnaissance aircraft on October 17th 1947. It was with No.812 Squadron from April 1948 until July 1948 moving on to No.737 Squadron at Eglinton from October 1949 until May 25th 1951. It moved to Ringway on November 9th 1951 for conversion to target tug status by Fairey's. It was delivered to Anthorn on January 21st 1953, joining No.700 Squadron at Ford, where it became '507/FD' in the July of the following year. It then changed its code to '503/FD' and from November 1955 to '506/FD'. It remained in use with No.700 until at least January 1957. By July it had joined the Fleet Requirements Unit at Hurn and remained with them until January 1958. In 1958 it was given official notification to be reduced to spares, but for some reason lingered long enough for this order to be cancelled in December 1962. It was moved to Culdrose and kept at the base for historical purposes in reserve for the Fleet Air Arm Museum.

Fairey Type Q Gannet AEW. Mk.3 XL503 '070/E'

Built 1960, constructor's number F.9461, three seat airborne early warning aircraft, powered by Bristol Siddeley Double Mamba 102 coupled turboprop of 3,875 horse power.

XL503 served for most of its operational life with No.849 Squadron 'D' Flight based at Lossiemouth or on H.M.S. 'Eagle'. It was coded '070/E' and still wears these markings today. It was retired by late 1970 and held in reserve with the Naval Aircraft Servicing Unit at Lossiemouth. It came out of storage in early 1972 and joined the Headquarters Flight for a while. By July 1972 it was with the Royal Radar Establishment at Pershore for work in connection with the AEW Shackletons. It was presented to the Museum in 1973.

Fairey Type Q Gannet COD. Mk.4 XA454 '264/H'

Built 1957, carrier on-board delivery aircraft, powered by one Bristol Siddeley Double Mamba 101 coupled turboprop of 3,035 horse power.

This aircraft was built as a standard anti-submarine AS. Mk.4 version. It first served with No.814 Squadron from May 1958, as '281/E'. It was at Culdrose by February 1960 in readiness for joining No.849 Squadron in March 1961, wearing the code '447'. It was then converted to COD Mk.4 status by removing most of the operational equipment. It joined the Flag Officer Aircraft Carriers on H.M.S. 'Victorious' from September 1964 in navy blue colour. It served until July 1965, by which time it was wearing the 'code' 'FOAC-V'. It was delivered to the Naval Aircraft Radio Installation Unit at Lee-on-Solent in June 1967. From October 1968 it was on board H.M.S. 'Hermes' coded '3'. It joined No.849 Squadron 'A' Flight from June 1970, taking on the codings it displays today '264/H'.

Fairey Type Q Gannet T. Mk.5 XG883 '773/BY'

Built 1957, two seat conversion trainer, powerplant as per XA454.

XG883 first flew on March 1st 1957. Its early service usage is not known, it being with No.849 Squadron as '451/CU' in May 1964. By September 1966 it was with the Headquarters Flight of the Squadron and remained as such, coded '773/BY', until September 1968. It joined the Museum in these colours shortly afterwards.

Gloster Meteor TT. Mk.20 WM292 '841'

Converted 1957, two seat target tug, powered by two Rolls-Royce Derwent 8s of 3,600 lb. s. t.

WM292 was originally built as an NF. Mk.11 night fighter. It was delivered to Armstrong

Whitworth at Baginton on August 23rd 1957 from No.5 Maintenance Unit at Kemble for conversion to target tug status. It was delivered, completed, to the Aircraft Holding Unit at Lossiemouth on April 24th 1958, being delivered to the Fleet Requirements Unit at Hurn in May, receiving the code 'O41'. It was back at Lossiemouth briefly in 1959, but rejoined the FRU in June 1959. On March 24th 1961 it was delivered to the Royal Naval Aircraft Yard at Sydenham for overhaul and was issued out to the AHU at Brawdy on August 15th 1961. It was taken on again by the FRU, again as 'O41', on October 27th 1961. It was delivered to Lee-on-Solent for modifications and returned to the FRU on September 9th 1966, this time taking up the code '841' by January 1969. It was delivered from Hurn to the Museum on June 4th 1969.

Grumman Martlet I AL246

Built 1940, single seat naval fighter, powered by one Wright Cyclone G-205A of 1,200 h.p.

This aircraft was an early Martlet (American designation F4F) without folding wings. It was delivered to No.802 Squadron at Donibristle in November 1940 and served with them until at least March 1941. From June 1943 it was with No.768 Squadron at Machrihanish until January 1944. Post-War it was bought by Loughborough College for use as an instructional airframe. It was acquired by the Navy in the mid 1960s and was an exhibit with the Museum from its opening.

Grumman Avenger AS. Mk.6 XB446 '992/CU'

Built 1940s, three seat electronic counter-measures aircraft, powered by one Wright Cyclone R-2600-20 radial piston engine of 1,750 horse power.

XB446 first served with the United States Navy as 69502, its history with the USN being unknown. It arrived on board a ship at King George V dock in Glasgow on May 15th 1953 and was taken to Royal Naval Air Station Abbotsinch. The aircraft was still in basic TBM-3E standard, so it was taken by road to Scottish Aviation at Prestwick for conversion to AS. Mk.4 configuration on January 13th 1954. It was delivered to No.820 Squadron at Eglinton on March 4th 1954, here it was coded '372/C'. It moved to the Naval Aircraft Radio Installation Unit at Gosport on March 3rd 1955 for conversion to AS. Mk.6 standard. The conversion completed, it was taken to Lee-on-Solent for flight testing on March 5th 1956. It remained there for further checks, maintenance and test flights until March 13th 1957 when it was delivered to No.751 Squadron based at Watton. Its starboard wing was damaged on November 28th 1957 and it was taken to Culdrose by road on December 3rd 1957, repairs being completed on January 21st 1958. It re-joined No.751 Squadron, now at Culdrose, on January 24th 1958. No.751 was redesignated No.831 Squadron on the 29th and XB446 was recoded '383'. During 1960 it was recoded '267' and it finally joined the Station Flight at Culdrose, as '992' on November 29th 1960. It was officially handed over to the Museum on September 4th 1969.

Hawker Sea Fury FB. Mk.11 WJ231 '115/O'

see photo 29

Built 1950, single seat fighter bomber, powered by one Bristol Centaurus 18 radial piston engine of 2,480 horse power.

First delivered to Royal Naval Air Station Anthorn in 1950 WJ231's history is not known at present. It was painted at one time inside the Museum with the serial WE726 to represent an aircraft from No.802 Squadron, but this has since been replaced by the correct serial. The codings worn are spurious. This aircraft has donated some parts to the Fleet Air Arm Historic Aircraft Flight's TF956 (see below).

Hawker Sea Hawk FGA. Mk.6 WV856 '163'

Built 1954, single seat naval ground attack fighter, powered by one Rolls-Royce Nene 103 turbojet of 5,400 pounds static thrust.

Built at Baginton by Armstrong Whitworth, WV856 was taken on charge at Abbotsinch on December 15th 1954. It was at the Royal Naval Aircraft Yard Sydenham in January 1955 for preparation for service with No.806 Squadron from April 1955 until July 1955, as '166/C'. It was at Anthorn in January 1956 and at Abbotsinch by July 1956. It was on a tour of the Mediterranean from January 1957. WV856 was moved to Fleetlands and was converted from an FGA. Mk.4 into FGA. Mk.6 status from July 1957 until January 1958. It then joined No.806 Squadron as '182/E' from May 1958 until July 1959. It was at Abbotsinch in January 1960 and on H.M.S. 'Centaur' the same month. It was in store at Fleetlands by January 1961 and moved to Sydenham in July. The following July it joined No.781 Squadron and served until September 1966. It was taken on charge with the Royal Aircraft Establishment on September 30th 1966. It is reported to have been scrapped in 1967, but was saved.

Hiller HT. Mk.2 XB480 '537'

Built 1953, basic trainer, powered by one Franklin 6V4-178 piston engine of 178 horse power.

Taken on charge at Gosport on July 13th 1953 and stored prior to issue to No.705 Squadron on January 9th 1954. It was moved to Lee-on-Solent on October 25th 1955 and entered storage on November 21st. It re-joined No.705 Squadron on May 28th 1957, as '540'. It was again put into store, this time at Donibristle, from April 24th 1958, moving to Fleetlands on January 19th 1959. It was back with the helicopter training unit, No.705 Squadron on October 10th 1960 and gained its current code '537'. It was retired on March 7th 1963 and moved to the Royal Naval Engineering College at Manadon, where it was given the maintenance number A2577. It was refurbished and later presented to the Museum.

Percival P.50 Sea Prince C. Mk.1 WF137

Built 1950, constructor's number P.50/5, communications aircraft, powered by two Alvis Leonides 125 radial piston engines of 550 horse power each.

WF137 was built at Luton and taken on charge by the Royal Navy on July 22nd 1950 at Lossiemouth. It was with No.781 Squadron from October through to January 1952. It was modernised by Field Aircraft at Tollerton from June 4th 1954 until July 1956. It then returned to service, with the Station Flight at Arbroath from July 1957, moving on to Yeovilton Station Flight on May 26th 1958. From June 6th it entered service with the Ferry Flight at Rochester. On September 18th it moved to the Aircraft Holding Unit at Lossiemouth, and it went back to Huntings at Luton in January 1959 for modifications. It then moved to Field Aircraft again for overhaul, this time at Wymeswold, from May 27th until October 10th, when it was delivered to the AHU at Brawdy. It was back with the Ferry Flight in July 1963 and on to Culdrose Station Flight by January 1964, coded '999/CU'. It flew into Yeovilton on December 6th 1968 to join the Museum.

Percival P.66 Sea Prince C. Mk.2 WJ350

Built 1953. communications aircraft, powerplants as above.

Taken on charge at Stretton on September 3rd 1953 WJ350 entered service in May 1954, probably with No.781 Squadron. By July 1956 it was the mount of the Flag Officer Naval Flying Training. It moved to Culdrose on June 3rd 1965 and continued in its previous role, that of a communications vehicle until its retirement. It moved to the Royal Naval Aircraft Yard for overhaul in January 1966, joining the Lossiemouth Station Flight from June and Yeovilton Station Flight from August 19th 1966. It was away for a while having a radio fit at the Naval Aircraft Radio Installation Unit at Lee-on-Solent, returning to the Yeovilton Station Flight on May 16th 1968. It joined the Museum on July 8th 1971. It was placed on display immediately, and serves as a good comparison with the C. Mk.1 above. It is hoped to put a Sea Prince T. Mk.1 crew-trainer on display on their retirement.

Saunders-Roe P.531 XN332 '759'

Built 1958, constructor's number S2/5268, prototype light helicopter, powered by one Blackburn Turbomeca Turmo 600 turboshaft of 325 estimated shaft horse power.

This aircraft was the second prototype P.531, taking to the air for the first time from Eastleigh on September 30th 1958 as G-APNV. It was transferred to the Royal Navy for trials in October 1959, and was based at Yeovilton with No.700X Flight as '759' until June 1961. It was relegated to instructional work and was issued to the Royal Naval Engineering College at Manadon, as A2579, before joining the Museum.

Sopwith Baby 'N2078'

Contemporary replica, using original parts, single seat floatplane, one Clerget rotary engine.

This aircraft was constructed from the parts of two Babies that were acquired by the now defunct Nash Collection. They have been built into 'The Jabberwok' given the serial 'N2078' - this aircraft was built by the Blackburn Aircraft and Motor Company in the original. The parts used came from 8214 and 8215, both built by Sopwith in 1915.

Supermarine Walrus I L2301

see photo 30

Built 1939, line number C3/21840, air sea rescue amphibian, powered by one Bristol Pegasus VI radial piston engine of 775 horse power driving a pusher propeller.

L2301 served only for a short time with the Royal A/F , being sold in 1939 to the Irish Air Corps, serialled N18. It was later sold to Aer Lingus as EI-ACC, but this was not taken up. In November 1946 it was bought by Wing Commander R. G. Kellett on behalf of No.615 (County of Surrey) Squadron to help in the recruiting of ground personnel. It cost £150. It was ferried from Dublin to Biggin Hill in March 1947 and given the registration G-AIZG. After some use as a joy-riding aircraft, it was sold for scrap in 1949. It was discovered in a very derelict condition in a scrap-yard at Thame by members of the now defunct Historic Aircraft Preservation Society. It was donated to the Fleet Air Arm and moved to Arbroath for complete refurbishing to its original wartime condition. It was officially handed over to the Museum on December 6th 1966.

Supermarine Seafire XVII SX137

Built 1945, single seat naval fighter, powered by one Rolls-Royce Griffon VI of 1,850 h.p.

SX137 was first at Royal Naval Air Station Culham, issued there on September 25th 1945. Until 1951 it was not with a unit as such, but with the following Stations or Aircraft Yards; Henstridge from January 28th 1946, Fleetlands from April 5th, Westlands for overhaul from June 17th 1947, Gosport from August 26th, Fleetlands from February 27th 1948, Anthorn from April 21st and Stretton from February 21st 1950. On June 18th 1951 it joined No.1831 Squadron Royal Navy Volunteer Reserve at Stretton, moving on to No.759 Squadron on November 6th. It then started another round of being attached to a Station; Stretton from May 12th 1953, Fleetlands from June 30th, Stretton from May 18th 1954, and Yeovilton from July 9th. It was struck off charge officially on May 6th 1953. In June 1962 it suffered the indignity of appearing as a 'Spitfire' at Earl's Court, with the serial 'W9132' and the name 'Suzy'. It moved to the Museum in April 1963.

Supermarine Attacker F. Mk.1 WA473 '146'

Built 1951, single seat naval fighter, one Rolls-Royce Nene 3 turbojet of 5,100 lbs. s. t.

This machine is the only Attacker left in existence. It was taken on charge at Culdrose on July 25th 1951. It first entered service with No.800 Squadron from October 1951 until March 1952, coded '102/J'. It moved then to No.702 Squadron and was with No.736 Squadron as '116/LM' after that. It was with Vickers briefly at South Marston before going to Abbotsinch on May 6th 1953. From November 1954 it was with Airwork General Trading, at Gatwick, until January 1955. It then cropped up again with Airwork G. T. in July 1955, but not at Gatwick, staying with them until July 1956. By September 1956 it was with Airwork at Lasham. It was retired and moved to Abbotsinch in January 1957 and was put on display at the gate as '146/J', it stayed here until at least June 1959. It was officially struck off charge on May 14th 1958. It went to the Museum in 1961.

Supermarine Scimitar F. Mk.1 XD317 '112/R'

Built 1959, single seat naval fighter, powered by two Rolls-Royce Avon 202 turbojets of 11,250 pounds static thrust.

XD317 was delivered to the Aircraft Holding Unit at Lossiemouth on November 11th 1959. It entered service on February 23rd 1960 with No.807 Squadron, as '197/R', until March 1961 when it was recoded '197/C'. It returned to the AHU on June 23rd 1961 and was delivered to the Royal Naval Aircraft Yard at Fleetlands on October 20th 1961. It was on strength at the AHU at Tengah, Singapore, from November 3rd 1961 and had joined No.807 Squadron again on February 12th 1962, this time as '195/C'. It was back at Lossiemouth on June 24th 1962, and at Fleetlands on May 16th ready for issue to No.736 Squadron on November 28th 1962. On January 14th 1963, it joined No.800 Squadron as '112/R' - the markings it wears today. It returned to Fleetlands on February 27th 1964 and from there to Lee on November 22nd 1965. Its role now changed to research and it joined the Royal Aircraft Establishment at Thurleigh from January 6th 1966, at Farnborough from March 10th 1966 and back at Thurleigh from April 1st 1966. After this it was flown to the Naval Aircraft Servicing Unit at Brawdy on July 27th ready for issue to the Fleet Requirements Unit. This happened on November 2nd 1966 when it was delivered to Hurn, becoming '033', recoded '833' by 1969. It returned to the RAE and Farnborough on August 4th 1969 and was delivered by road to the Museum in September 1969 at the end of its flying life.

Westland WS-51 Dragonfly HR. Mk.5 WN493

Built 1953, constructor's number WA/H/70, naval helicopter, powered by one Alvis Leonides.

WN493 first flew, as a HR. Mk.3, on March 5th 1953. It was delivered to Gosport twenty days later. It was transferred to the Aeroplane and Armament Experimental Establishment on June 12th at Boscombe Down for trials. These trials are thought to have included flights off H.M.S. 'Perseus'. It was back at Gosport on June 2nd 1954 and was issued out to the local No.705 Squadron on September 3rd, coded '705'. It moved to Donibristle on July 4th 1955 for reconditioning, rejoining No.705, via Lee-on-Solent, on June 12th 1956, this time as '534'. It returned to Donibristle on December 9th 1957 for conversion to HR. Mk.5 standards. The conversion was finished on February 1st 1958, going back to No.705 on April 2nd, with no traceable code. It was briefly with No.701 Squadron from April to September 1958 before re-joining No.705, now at Lee-on-Solent. In 1959 the unit moved to Culdrose, taking WN493 with it. It was reconditioned at Fleetlands from September 27th 1960 and returned to No.705 Squadron on July 9th 1962, probably coded '549'. It was later transferred to the Culdrose Station Flight, and was coded '915/CU'. It was transferred to the care of the Museum on October 14th 1966.

Westland W.34 Wyvern TF. Mk.1 VR137

Built 1947, single seat torpedo fighter prototype, one Rolls-Royce Eagle of 2,690 h.p.

This aircraft was completed at Yeovil, but is reported to have never been flown, the Eagle having been superceded as the powerplant for the Wyvern. It was acquired by the College of Aeronautics at Cranfield and moved there by road on November 30th 1950. It remained at Cranfield until purchased by the now defunct Historic Aircraft Preservation Society in May 1963. It was loaned to the Museum, and later became their property. It was delivered to Yeovilton in February 1966. It was painted up to represent a standard service Wyvern in the winter of 1970/71.

Westland WS-55 Whirlwind HAR. Mk.1 XA864

Built 1953, general purpose naval helicopter, powered by one Alvis Leonides Major 755.

XA864 first flew, with the test registration G-17-1, at Yeovil on August 29th 1953. It was officially transferred to Westlands for trials from March 31st 1954. Westlands used it for power control and resonance tests and for trials with the Vokes exhaust silencer. It was taken on charge by the Navy on April 2nd 1954. By February 22nd 1955 it was back with Westlands, and may have been also with the Royal Aircraft Establishment at Farnborough in 1956. On October 17th 1957 it was delivered to the Aeroplane and Armament Experimental Establishment at Boscombe Down, and from January 13th 1959 with the RAE at Farnborough. On February 3rd 1965 it was moved to the RAE at Thurleigh and used by the Radio Flight for general work. It is also reported to have been used in some capacity by the Kestrel Tripartite Evaluation Squadron. It was accepted by the Museum on March 3rd 1970.

Notes : The following incomplete airframes are on display :-

Blackburn Skua I L2940 remains recovered from Lake Grotli in Norway. Ex No.800 Squadron, downed on April 27th 1940, returned to Lee-on-Solent August 27th 1974.
De Havilland DH.115 Sea Vampire T. Mk.22 XA127, fuselage pod only.
Fairey Barracuda II DP872 Boulton Paul built, remains from Ulster.
Gloster Sea Gladiator II N5903 remains only, ex Shuttleworth Collection. (More details can be found in the entry for the Trust's Gladiator)
Short 184 8359 Built by Westland, on loan from the Imperial War Museum. Forward fuselage only, delivered from Duxford on January 28th 1976.

The Museum has several aircraft in store, on site at Yeovilton, and elsewhere, and these will be added to the exhibition when space permits. These are list below.

The following aircraft are held in reserve for future display :-

Bensen B.8M G-AZAZ, built by RNEC Manadon, c/no. RNEC.1	kept at Yeovilton.
De Havilland DH.112 Sea Venom FAW. Mk.22 XG737 '438/BY'	kept at Yeovilton.
Fairey Albacore I N4172 remains found at Tongue.	kept at Yeovilton.
Fairey Type Q Gannet T. Mk.2 XA508 '627/GN' from Wroughton by road 11/9/75	kept at Yeovilton.
Fokker Dr.1 replica '102/17' based on Lawrence Parasol ex Singapore	kept at Yeovilton.
Gloster Meteor T. Mk.7 WS103 gloss black and dayglo scheme	kept at Wroughton.
Hawker Sea Fury FB. Mk.11 VR930 from St. Athan and Colerne previously	kept at Yeovilton.
Hawker Sea Hawk F. Mk.1 WF219 in sections, A2439	kept at Yeovilton.
Hawker Sea Hawk FGA. Mk.6 XE340 '131/Z' ex Staverton and Brawdy gate	kept at Wroughton.
Saunders-Roe P.531 XN334 A2525, ex No.700X Flight	kept at Yeovilton.
Supermarine Scimitar F. Mk.1 XD220	kept at Wroughton.
Westland WS-51 Dragonfly HR. Mk.1 VZ965 reported to be from Cyprus	kept at Culdrose.
Westland WS-51 Dragonfly HR. Mk.3 WG718	kept at Yeovilton.
Westland WS-55 Whirlwind HAS.Mk.3 XJ402 A2572/61. From Manadon 13/ 2/76	kept at Yeovilton.
Westland WS-55 Whirlwind HAS.Mk.3 XG757 A2575, ex Lee-on-Solent	kept at Wroughton.
Westland WS-55 Whirlwind HAS.Mk.7 XG594 '517'	kept at Wroughton.

The Fleet Air Arm Historic Aircraft Flight is also based at Yeovilton. Its aircraft are not available for public inspection at Yeovilton, but appear at many air shows.

De Havilland DH.82A Tiger Moth II T8191 built by Morris Motors at Cowley, 1940, flies in

1940 style camouflage. Totally refurbished winter of 1976/77.
Fairey Swordfish II LS326 '5A' built by Blackburn Aircraft at Brough in 1943. Civil post-war with Fairey Aviation as G-AJVH. Presented to the Fleet Air Arm in 1959.
Fairey Firefly AS. Mk.5 WB271 '204/R' constructor's number F.8497, now flies with a Griffon 57, first flew as such on March 2nd 1976.
Hawker Sea Fury FB. Mk.11 TF956 '123/T' first production FB.11. Allocated G-9-395 by Hawkers after purchase from FAA. Gift to the FAA from HSA. *See photo No.29.*
Hawker Sea Fury T. Mk.20 WG655 flown in from Lubeck 29/ 6/76 as D-CACU. Ex FAA airframe reconditioned to target tug for German Government, G-9-65, later ES.3616. For 1977 season.

GLASGOW MUSEUM OF TRANSPORT

Glasgow Museum of Transport contains one of the most comprehensive collections of historic transport items in the United Kingdom.

The Museum has been open since 1964, when six ex-Glasgow Corporation tramcars together with a number of early motor cars, motor cycles and bicycles were placed on exhibition. The display has since been widened in scope to include most forms of mechanical transport, and now covers more than twice its original area. Seven railway locomotives, representing each of the five major Scottish companies at the grouping of 1923, have been installed in a large extension, and more recent additions are new galleries for the interesting collection of horse drawn and commercial vehicles and fire engines.

Two aircraft are on display in the Museum. The technology gallery at Kelvingrove, Glasgow's principal museum, contains full-sized sectioned examples of the Rolls-Royce Merlin Mk.XX and Dart Mark V aero engines.

It is hoped that the extensive collection of ship models, at present also at Kelvingrove, will be re-housed in a large new gallery at the Museum of Transport during 1977.

The Museum is situated on the south side of Glasgow, and is easily reached by car or train. Albert Drive runs westwards from the A77 road to Kilmarnock, about one mile south of the city centre and five minutes drive from the M8 motorway. Street parking is usually to be found within a few minutes walk. Pollokshields East Station, adjacent the Museum, is served by frequent local services from Glasgow Central.

The Museum is open from 1000 to 1700 hours Monday to Saturday and 1400 to 1700 hours on Sunday. Admission is free, and there is a tea-room and souvenir shop.

Address: Glasgow Museum of Transport, 25, Albert Drive, Glasgow, G41 2PE.
Telephone : 041-423-8000.

Kay Gyroplane Type 33/1 G-ACVA

Built 1934, constructor's number 1002, experimental single seat autogyro, powered by one Pobjoy R radial piston engine of 75 horse power.

This aircraft was designed by David Kay in the early 1930s and was built by Oddie, Bradbury and Cull Limited at Eastleigh in 1934/35. It first flew on February 18th 1935 and spent a while on test flying and trials. Nothing ever came of the design and G-ACVA is last known to have flown on August 16th 1947 at Scone, having been stored through the War. It was stored at Scone until being presented to the Museum in 1969.

Pilcher Hawk replica BAPC.48

Replica built in the 1960s, following original lines.

This is a replica of the famous hang-glider of 1896, in which Scot Sir Percy Pilcher made many flights. It was built by the cadets of No.2175 Squadron Air Training Corps and presented to the Museum. It has been allocated the BAPC identification number 48.

HISTORIC AIRCRAFT MUSEUM

The Historic Aircraft Museum is the result of considerable effort to establish a museum at Southend.

There is an indoor and an outdoor display area, housed in very modern surroundings. During the winter season exhibits are moved about for repairs, refurbishing etc. Thanks to the combined efforts of the staff and members of the local Vintage Aircraft Society the exhibits are kept in first class condition and one must remember that many of the aircraft now on display were on their way to the scrap heap.

On entry to the main hall there is a cafeteria to the left, capable of seating 120 persons at one sitting and on the right the 'Kit Room', which is really a shop full of aircraft kits, souvenirs, books, postcards etc. The static exhibition hall contains World

28 *The FAAM's Fokker Dr.I at Seletar* P. R. March

29 *Sea Furies TF956 and WJ231 at Hurn 10/8/61* P. H. Butler

30 *Walrus L2301 just after restoration*

31 *Gadfly G-AVKE at Southend* Avia Press

32 *Javelin XH768 as 'E' of 25 Squadron* Roger Lindsay

33 *Lincoln RF342 at a very wet Cranfield* P. H. Butler

34 *Camel 'F3043' arrives back at the IWM from the Royal Tournament* *IWM MH 3968*

35 *Camel, Bristol Fighter and Bleriot at the Royal Tournament. Imperial War Museum MH 3967*

36 *Loading the Bristol Fighter into the aviation gallery, Lambeth* *IWM MH 3981*

War Two components, aero engines and such modern items as the wheel and brake assembly off the first Concorde.

In the main hall are housed many of the 'rare birds' that make the Museum so interesting, the replica of the 1909 Triplane by A. V. Roe, that famous collection of De Havilland Moths, the famous D.H. Drover from Australia and, of course, the pride of the Museum - the D.H.84 Dragon, together with the Flying Fleas, Sea Fury and many others, many of these are in full flying condition and can be seen at many air displays during the season.

On the outside is housed a full range of jet aircraft from the past which includes such items as the Gloster Javelin and Meteor, Sea Hawk, Saab, F-84F together with German aircraft such as the Heinkel and the Storch. The giant Blackburn Beverley is open to the public as are many other aircraft. There is a large group of helicopters, the largest being the Widgeon.

The motto of the Museum is that of public participation and we try to keep a minimum of 'do not touch' notices.

Until June 6th 1977 the Museum will be open on Saturdays and Sundays only from 1100 to 1800 hours. The Museum will resume normal opening hours from 1000 to 1800 hours seven days a week with effect from Monday June 6th 1977. Group visits can be arranged outside of normal opening hours, but on Mondays only during the winter months.

Admission is as follows : Adults 35p, Children 15p and Old Age Pensioners 10p.

The Museum is located on the southern side of Southend Airport.

Address : The Historic Aircraft Museum, Aviation Way, Southend-on-Sea, Essex.

Telephone : 0702-546344.

Avro 652A Nineteen Series 2 G-AGPG

Built 1945, constructor's number 1212, avionics test-bed, powered by two Armstrong Siddeley Cheetah 15 radial piston engines of 420 horse power each.

G-AGPG was built at Woodford and registered to the Avro Company as such on August 6th 1945. In 1952 it was converted from its previous Series 1 standard to Series 2 guise. It was in use with several concerns, but the most famous took over ownership in October 1967. That was the Ekco Electronics Limited, who based G-AGPG at Rochford and used the craft for general communications, but more especially for radar and other avionics testing. It was fitted with a bulbous nose to accomodate a wide range of possible radar installations. In April 1969 G-AGPG was registered to Pye Telecommunications, who had absorbed the Ekco Company. Upon retirement it was presented by the Company to the Museum in immaculate condition.

Avro 694 Lincoln II G-29-1

see photo 33

Built 1946, aerodynamic research test-bed, powered by four Rolls-Royce Merlin 68s of 1,635 horse power each.

This aircraft served with the Royal Air Force as a Lincoln B. Mk.2, serial RF342, before being sold to D. Napier and Son Limited, at Luton, in January 1959. It was registered as G-APRJ and was prepared for use by that company into research into the icing-up of airframes at height. From 1960 it adopted the 'B' class registration G-29-1 which allowed it to undertake more adventurous testing. In order to keep G-29-1 flying, Napier's dismantled Lincoln G-APRP (previously RF402) at Luton in 1960 to supply spares. In November 1962 it was sold to the College of Aeronautics at Cranfield and continued research in a similar manner to Napier's. With the College G-29-1 became G-36-3. It was acquired by the British Historic Aircraft Museum (the abortive predecessor of the HAM) and flown into Rochford on May 9th 1967 as G-APRJ. With the opening of the HAM in 1971 it had been repainted to represent its days with Napier at Luton.

Blackburn B.101 Beverley C. Mk.1 XB261

see photo 26

Built 1955, long range large capacity transport, powered by four Bristol Centaurus 173 radial piston engines of 2,850 horse power each.

XB261 first flew from Brough on July 5th 1955, and was officially awaiting collection by the Royal Air Force on the 20th. It was never to serve with the RAF, spending its entire operating life with the Aeroplane and Armament Experimental Establishment at Boscombe Down, on a variety of tasks and trials. Its main duties with the A&AEE involved using it as a stable and large-capacity parachute platform. During 1966/67 it was used to develop the lo-drop system of supplying trops, using a low flying aircraft. This developed into the ULLA system, Ultra Low Level Air-drop. This was a system of dropping pallatised loads (including Land Rovers and field guns) from an aircraft flying at low level, by streaming a number of parachutes out from the rear of the aircraft and pulling the load out of the hold.

It was retired and flown into Rochford on October 6th 1971 from Boscombe Down.

CASA 2.111 '6J+PR'

Built 1951, constructor's number 049, medium bomber, powered by two Rolls-Royce Merlin 500 piston engines of 1,600 horse power each.

This is a licence-built Heinkel He 111H-16 built by Construcciones Aeronauticas S.A. for the Spanish Air Force, serial number B2-157. It was completed on December 3rd 1951 and used by the SAF for many years, latterly for target towing and other second-line work. It was sold in the United Kingdom in 1968 and registered to Spitfire Productions Limited as G-AWHB for use in the film 'The Battle of Britain'. It was ferried from Madrid to Duxford via Jersey on May 21st 1968. After the film it was stored at West Malling until acquired by the Museum and installed at Southend in its film colours by March 1972.

De Havilland DH.60G Gipsy Moth G-ATBL

Built 1933, constructor's number 1917, basic trainer, one De Havilland Gipsy I of 100 h.p.

Built in 1933 at Stag Lane to an order from Switzerland, this aircraft was delivered, as CH-353, in March. It was later re-registered, when the new system was introduced, as HB-OBA. It was imported into the United Kingdom in February 1965, and given the registration G-ATBL, which at first seems totally out of keeping with its age. It was taken over by Chris Lovell in November 1969 and restored to the immaculate condition in which it stands to date. It became the property of Anthony Haig-Thomas is May 1972 and is fully airworthy.

De Havilland DH.80A Puss Moth G-AEOA

Built 1931, constructor's number 2184, touring monoplane, powered by one De Havilland Gipsy Major in-line piston engine of 130 horse power.

This aircraft was built for an order in Yugoslavia at Stag Lane and was despatched, as UN-PAX in July 1931. It was later re-registered as YU-PAX and imported into the United Kingdom in October 1936 by Airwork Limited, who registered it as G-AEOA. It was impressed on May 31st 1941 as ES921 and used by the Headquarters Air Transport Auxiliary at White Waltham as a crew ferry and general 'hack'. It was at No.5 Maintenance Unit at Kemble on May 3rd 1945 and put up for disposal. It was restored to East Anglian Flying Services on February 1st 1946. In September 1946 it was registered to R. L. Whyham and, from December, to Whyham and Bellamy Air Transport, followed by Air Navigation and Trading Limited from October 1947, all based at Squires Gate. By February 1949 it was flying with the Speedbird Flying Club, later renamed Airways Aero Associations Limited. It made its last flight on December 17th 1949 and was delivered to Squires Gate by road on December 17th 1949 for long term storage. It became the property of ANT again in August 1952, but was to move considerably around the country, in the way that many apparently immobile aircraft do. It moved to Hurn, then to Romsey and finally to the grounds of Hants and Southampton University for rebuild. This was supervised by Dr. J. H. B. Urmston, to whom it was restored on June 5th 1967. It first flew again at Booker on September 11th 1968. It passed on to Mr. Haig-Thomas on August 27th 1973 and is fully airworthy.

De Havilland DH.82A Tiger Moth DE-208/G-AGYU

Built 1942, constructor's number 85265, basic trainer, one De Havilland Gipsy Major I.

This machine was built for the Royal Air Force as DE208 and was officially taken on charge at No.46 Maintenance Unit at Lossiemouth on January 2nd 1942. From July 12th 1942 it was with the Station Flight of Holme-on-Spalding-Moor, followed by a spell at Wickenby from May 25th 1943. It then joined No.3 Elementary Flying Training School at Shellingford on July 21st 1944, suffering a Category B (Beyond repair on site) flying accident on September 19th 1945. It was delivered to Marshalls of Cambridge at Teversham on January 6th 1946 and registered to them four days later as G-AGYU. It flew from April 1946 until February 1952 with Marshalls, moving only as far as the Cambridge Aero Club on the same airfield until May 1966. On that date it became the property of the Essex and Suffolk Gliding Club and earned its keep towing gliders aloft until becoming part of the Haig-Thomas Moth Collection on June 23rd 1971, and being repainted into its war-time markings. It is airworthy.

De Havilland DH.82A Tiger Moth G-ANPE

Built 1940, constructor's number 83738, role and powerplant as above.

G-ANPE had a very ignominious start in life as T7397 on October 5th 1940 when it was placed straight into store at Webbs Garage, Banbury Road, Bicester. Not until July 28th 1941 did it enter service with No.18 Elementary Flying Training School at Fairoaks, it stayed with this unit until July 30th 1945 when it moved to No.10 Maintenance Unit at Hullavington for storage, moving to No.12 MU at Kirkbride on August 13th 1951. It re-entered service with No.19 Reserve Flying School on June 24th 1952, returning to No.12 MU on July 31st 1953. It was sold off in March 1954 and it was registered as G-ANPE to R. A. Peacock (Aviation) Ltd. It was taken to Croydon and converted to French standards and registered as F-BHAT, being first used by the Aero Club du Jura. It was re-imported into the United Kingdom by M. J. Coburn and C. C. Hughes in September 1970 and restored as G-ANPE. It joined Budge Brothers in October 1971 and remained with them until their demise, and is now the property of the Museum.

De Havilland DH.83 Fox Moth G-ACEJ

Built 1933, constructor's number 4069, four or five seat feederliner/air taxi aircraft, powered by one De Havilland Gipsy Major I of 130 horse power.

Built to the order of the Scottish Motor Traction Company Limited at Renfrew and delivered to them in May 1933 as G-ACEJ. From 1936 it took up its famed role of joy-rider from the sands of Birkdale, near Southport, with Giro Aviation. It being officially registered to them on July 22nd 1936. It was stored through the war and served on from the beach until bought by Mr. N. H. Jones on June 16th 1966. It became the property of Anthony Haig-Thomas on October 26th 1972. In 1976 it was repainted in the immaculate gloss black and gold trim that it currently wears. It is fully airworthy.

De Havilland DH.84 Dragon I G-ACIT

see photo 27

Built 1933, constructor's number 6039, feederliner, powered by two D.H. Gipsy Majors.

The Museum's flag-ship, 'The Orcadian', flew into Southend on September 19th 1971 and remains in airworthy condition. It was built at Stag Lane and first flew in July 1933. It was delivered that month to Highland Airways Limited and was called 'Aberdeen'. In June 1938 it moved to Scottish Airways, who christened it 'Orcadian'. In June 1940 it was taken on by the Associated Airways Joint Committee, which managed internal airline routes in the United Kingdom during the War. From February 1947 it joined British European Airways, based at Dalcross for Scottish and Island use. In January 1952 it joined Air Navigation and Trading Limited and was to spend ten years mostly flying joy-riders around Blackpool Tower. It became the property of the ill-fated Beagle Aircraft at Rearsby from March 1962. It became the property of the Museum in June 1971.

De Havilland DH.87B Hornet Moth G-ADLY

Built 1935, constructor's number 8020, touring biplane, one De Havilland Gipsy Major I.

G-ADLY was built for W. Lindsay Everard as a DH.87A (tapered wing tips) and given the name 'Leicestershire Foxhound II'. It was subsequently converted to the square tipped DH.87B standard. It was impressed on February 28th 1940 as W9388 and joined No.4 Coastal Patrol Flight at Hooton Park the following day. This unit was renumbered No.3 CPF on March 1st 1940. It went to No.39 Maintenance Unit at Colerne on June 25th for painting into military colours, joining the Calibration Flight of No.2 Radio Maintenance Unit four days later. In October this unit became the No.2 Radio Servicing Section and later, in February 1941, No.71 (S) Wing. On June 15th 1943 it joined No.526 Squadron until going to No.5 MU at Kemble on February 16th 1946. It was restored on May 20th 1946 and was to be owned by a string of flyers, including the Viscountess Mairi Bury from January 1950, Viv Bellamy from March 1959 and Peter Clifford from March 1963. It was bought by Mr. Haig-Thomas on January 13th 1972, it is fully airworthy.

De Havilland DH.94 Moth Minor Coupe G-AFNG

Built 1939, constructor's number 94014, basic trainer, one De Havilland Gipsy Minor of 90 hp.

G-AFNG was built at Hatfield for the Cambridge Aero Club as an open cockpit version and was delivered to them in July 1939. It was impressed on June 9th 1940 as AW112 at No.45 Maintenance Unit at Kinloss. From July 2nd 1942 it was on charge at Binbrook, moving to the Station Flight at Wyton on March 20th 1943. It moved to No.5 MU at Kemble for maintenance on July 26th 1943 and remained in store there. On May 31st 1945 it re-entered service with the Empire Air Armament School at Manby, returning to No.5 MU for disposal on January 10th 1946. It was restored as G-AFNG to Mr. D. G. S. Cotter on May 30th 1946 and was converted to a Coupe in July 1953. By November 1958 it was in use with the Blackpool and Fylde Aero Club and was in several private hands until being bought by Mr. Haig-Thomas on June 6th 1968, until being sold in January 1969. Mr. Haig-Thomas bought the Coupe back on May 25th 1972 and it is still airworthy.

De Havilland DH.115 Vampire T. Mk.11 XK625 '12'

Built 1956, constructor's number 15782, advanced trainer, powered by one De Havilland

Goblin 35 turbojet of 3,500 pounds static thrust.

XK625 was built at Hawarden and was handed over to the Royal Air Force on July 25th 1956. The following day it was on charge with No.7 Flying Training School at Valley. It moved on to No.1 FTS at Linton-on-Ouse, coded '50', from January 29th 1958. It returned to Valley, where No.4 FTS was now the tenant, on August 21st 1962, and received the code '60'. Then came the turn of No.8 FTS at Swinderby, coded '12', the colours of which the exhibit still displays. It was flown to No.19 Maintenance Unit at St. Athan in August 1963 for disposal. It was bought by Hawker Siddeley Aviation on December 4th 1968 and moved into storage at Hawarden and later at Woodford. It was hoped to find an overseas customer, but this did not happen and XK625 was offered, along with several other Vampires, to member groups of the British Aircraft Preservation Council. It was delivered to the Museum by road on March 20th 1973.

De Havilland Australia DHA-3 Drover 2 'VH-FDT'

Built 1951, constructor's number 5014, eight seat 'bush' airliner, powered by three De Havilland Gipsy Major 10 Mark 2 in-line piston engines of 145 horse power each.

The Drover is a much-modified Dove and this machine was built at Sydney as VH-EAZ, later becoming VH-EAS. It was acquired by Air Navigation and Trading Limited in December 1960 and registered as G-APXX. It arrived at Liverpool Docks on September 19th 1961 and was delivered by road to Squires Gate. It was not used and was acquired by the current Museum's predecessors and moved to Southend on May 19th 1967. It has been acquired by the current Museum and painted in typical flying doctor colours, a role for which this type became famous.

Fairchild M-62A Cornell N9606H

Date of construction unknown, primary trainer, thought to be powered by one Ranger L-440-7 piston engine of 200 horse power.

Very little is known of this aircraft, except that it arrived in the United Kingdom in a crate in 1974, in very poor condition. It was reported to have been part of a swop for the dismantled Sea Fury of the Museum, but this has not been confirmed. It still remains in a damaged condition. Further details appreciated.

Fiat G.46-4 BAPC.79 'FHE'

Built 1948-51, constructor's number possibly 106, basic trainer, powered by one Alfa 115 in-line piston engine of 225 horse power.

The history of this aircraft is another blank. It was a gift from Italy, and once served with the Italian Air Force with the code 'FHE'. It was flown into Northolt in 1970 inside a DC-6 and later moved to Shoreham for storage. It joined the Museum in 1971 and was repainted in representative Italian Air Force colours.

Fiesler Fi 156C-3 Storch 'CB+VD'

Built 1943, constructor's number possibly 110064, communications aircraft, powered by one Argus As10C in-line piston engine of 200 horse power.

This aircraft was delivered to the Swedish Air Force in June 1943, it was serialled Fv3812 and given the Swedish designation S.14B. It was sold in Austria as OE-ADR and later joined Dianaflug at Hamburg as D-EKMU for banner-towing operations. It was acquired by the Museum in 1972 and is now painted up in typical Luftwaffe North African camouflage and called 'Max'.

Gloster Meteor T. Mk.7 VZ638

Built 1949, advanced trainer, powered by two Rolls-Royce Derwent 8s of 3,600 lbs s. t.

VZ638 was awaiting collection on July 28th 1949 and was delivered to No.500 Squadron the following day, suffering a Category 3 (Repair beyond unit capacity) flying accident on August 17th 1950. No.49 Maintenance Unit looked at it from August 25th but found it had to be returned to Gloster's on September 14th until October 11th. It re-joined No.500 Squadron on November 1st and served until April 1st 1951 when it joined No.25 Squadron. It then moved to No.54 Squadron on July 15th 1953 and then to No.85 Squadron from December 21st 1953. From February 25th 1954 it was on charge with the Fighter Command Communications Squadron until being involved in an accident on March 17th 1954. No.71 MU effected a repair on site from March 19th until May 31st. It moved to the Station Flight of Biggin Hill on June 11th 1954 for a mere ten days before joining No.501 Squadron. It flew to Tarrant Rushton on February 8th 1956 for a complete refurbish by Flight Refuelling Limited. This work was completed on May 31st and it joined the All Weather Operational Conversion Unit (No.237 OCU) on June 15th. It was retired to No.20 MU at Aston Down on May 8th 1958 and started a shuffle between MU's : No.33 at Lyneham from July 25th, No.5 at Kemble from July 6th 1961, No.33 again from August 11th. It re-entered service on February 1st 1962 with No.5 Flying Training School for just 26 days before joining No.8 FTS. It moved to Cranwell and the RAF College, coded 'X', from October 1st and again suffered an accident on September 26th 1963. No.60 MU from Dishforth took command this time. It returned to the unit, still as 'X', but now with a new name, the College of Air Warfare on December 11th. It was finally retired to No.5 MU on May 3rd 1965 and moved to the Museum on January 12th 1972.

Gloster Javelin FAW. Mk.9 XH768

see photo 32

Built 1957, all weather fighter, powered by two Bristol Siddeley Sapphires of 11,000 lbs s.t.

XH768 was built as an FAW. Mk.7 but was not used operationally as such, being stored at No.5 Maintenance Unit at Kemble. It returned to the makers at Hucclecote and was converted to FAW. Mk.9 status from March 1959 until September 1959. It was officially taken on charge on September 30th 1959 and issued to No.19 MU at St. Athan on October 5th. It joined No.25 Squadron, as 'E', on December 8th 1959. No.25 renumbered as No.11 Squadron on December 3rd 1962 and XH768 remained with them as 'E'. On January 11th 1966 it moved to serve with No.29 Squadron, this time as 'O' until being retired to No.27 MU Shawbury on November 9th 1966. It was moved to Cranwell and joined the Engineering Flight there as instructional airframe 7929M until acquired by the Museum.

Hawker Sea Fury FB. Mk.11 WJ288 '029'

Built 1949, single seat naval fighter, powered by one Bristol Centaurus 18 radial piston engine of 2,480 horse power.

WJ288 was taken on charge at Donibristle and later served at Anthorn, Lossiemouth and finally with the Fleet Requirements Unit as Hurn, coded '029'. It was struck off charge in 1962 and used by Hawker Siddeley at Dunsfold for spares for export Sea Fury orders. It had recorded a total flying time of 1,193 hours. After many months in the open it was delivered to the embryo museum at Biggin Hill in 1966 and removed to Southend by road on May 6th 1967. After much time in the open after the failure of the initial museum it was rescued by the present Museum and restored to its superb condition of today.

Hawker Sea Hawk FGA. Mk.6 'XE364' '485/J'

Built 1955, single seat naval ground attack fighter, powered by one Rolls-Royce Nene 103 turbojet of 5,400 pounds static thrust.

This aircraft is really XE489, but is painted to represent an aircraft that saw action from H.M.S. 'Eagle' with No.899 Squadron during the 1956 Suez Campaign. XE489 was taken on charge at Abbotsinch on January 10th 1956 and remained there until at least July. It was touring in the Mediterranean from October and was at Hal Far from January 1957 until July 1958, returning to the United Kingdom on December 1st 1958. By January 1959 it was up at Lossiemouth and back in the 'Med' from the 22nd, returning on March 12th 1959. Squadron usage at this time is not confirmed. It was at Brawdy in July 1959 and at Abbotsinch by January 1960, at Fleetlands by July and back at Abbotsinch by January 1961. It then joined the Fleet Requirements Unit at Hurn on June 17th and had become '026' by April 1963. It was overhauled at the Royal Naval Aircraft Yard at Sydenham from September 3rd 1963, returning to the FRU on October 11th 1963. It was acquired by the predecessors of the Museum on May 20th 1967 and moved to Southend. XE489 has a total of 1,402 flying hours.

Mignet HM.14 Pou du Ciel G-ADXS

Built 1935, constructor's number CLS.1, single seat ultra-light, powered by one Scott A.2S Flying Squirrel two-cylinder piston engine of about 25 horse power.

This aircraft has strong local connections being built at flown at Rochford by C. L. Story. It flew briefly from 1935, carrying the name 'The Fleeing Flea'. It went into store before the War and remained there until 1959 when it was refurbished and put on display in Southend. From 1964 it was displayed at the Skyfame Museum at Staverton until going on display with the Historic Aircraft Museum, on loan from the builder's family.

Mignet HM.293 G-AXPG

Built 1969 to 1972, constructor's number PFA.1333, single seat ultra-light, powered by one converted Volkswagen piston engine of 1,300 c.c.s.

This aircraft is a modern design 'Flea' and provides a good comparison with the original above. G-AXPG was built by Bill Cole at Hazeleigh Grange and it first flew in 1972. It is active during the summer months, and winters at the HAM.

Miles M.65 Gemini 1A G-AKGD

Built 1947, constructor's number 6492, light cabin twin, powered by two Blackburn Cirrus Minor 2 in-line piston engines of 130 horse power each.

This aircraft has little in its past of real note. It was registered on October 6th 1947 as G-AKGD and served for some twenty three years. It was last used as a chase-plane for the Vickers Vimy replica now on display in the Royal Air Force Museum. After this it fell into a state of disrepair, and was deemed to be beyond economical repair to flying condition. It was donated to the Museum in 1972 by 'Dizzie' Addicot.

North American Harvard IIB '93584' 'LTA-584'

Built 1943, constructor's number 14A-639, advanced trainer, powered by one Pratt and Whitney R-1340-49 radial piston engine of 550 horse power.

This aircraft was built for the United States Army Air Force as 42-12392 by the Noorduyn Aircraft Company in Canada. It was diverted to the Royal Canadian Air Force and taken on charge by them on March 23rd 1943 as FE905. It is assumed to have been in store for most of its life with the RCAF as its total flying time when it was struck off charge on November 27th 1946 was just 74 hours. It was stored until bought by the Danish Air Force and delivered to Kastrup in September 1949. It was officially handed over, as 31-329, on September 11th 1950. It was struck off Danish charge, by then as just 329, on August 31st 1960, with a total flying time of 1,903 hours. It was sold in Norway and used by Fjellfly at Skien as LN-BNM. It was acquired by the Museum and shipped in, via Felixstowe docks, in May 1972. It has been painted to represent a standard USAF AT-6.

North American TB-25J-NC Mitchell 'HD368' 'VO-A' see photo 50

Built 1944, constructor's number 34136, aerial camera platform, powered by two Wright R-2600-29 radial piston engines of 1,700 horse power each.

This aircraft is currently painted to represent a Mitchell III of No.98 Squadron, Royal Air Force. It was actually built as a standard B-25J medium bomber for the United States Army Air Force, serial number 44-30861. It was later converted to a trainer aircraft, and designated a TB-25J. It was disposed of and was bought for aerial filming work, being registered as N9089Z. The conversion included the modifying of the nose to accept an optically perfect plexiglass nose cone to facilitate filming, and the rear hatches were similarly treated. As such it arrived in the United Kingdom in 1963 and was involved in the filming of 'War Lover' and '633 Squadron'. After these two it was flown into Biggin Hill and slowly decayed. It was bought by the British Historic Aircraft Museum (the predecessors of the current Museum) in 1966 for the princely sum of £100. It was moved to Southend and spent much time in the open until the current Museum restored it to its present hue.

Percival Proctor IV NP303

Built 1944, line number H-687, cabin monoplane, powered by one De Havilland Gipsy Queen 2.

This aircraft was built for the Royal Air Force as NP303 by F. Hills and Sons at Trafford Park, Manchester. It was civilianized in January 1957 as G-ANZJ. It was acquired by the BHAM and flown into Rochford from Baginton in 1967. With the demise of the BHAM it was to remain outside until taken on by the Historic Aircraft Museum in 1971. By this time much of the structure had decayed and it was given the wings from G-AOBW (see below) as its own wings were beyond repair. At the same time it was painted in representative RAF markings.

Percival P.56 Provost T. Mk.1 WV483 'N-E'

Built 1953, constructor's number P56/46, basic trainer, powered by one Alvis Leonides 126 radial piston engine of 550 horse power.

WV483 was ready for collection by the RAF at Luton on September 30th 1953 and was delivered to No.6 Flying Training School on October 23rd, taking up the coding 'N-E'. These markings it wears today. It was retired to No.27 Maintenance Unit at Shawbury on July 16th 1957 and was officially declared non-effective stock on June 28th 1960. It was issued to No.4 School of Tecnical Training at St. Athan on October 25th as an instructional airframe, number 7693M. It was acquired by the Museum in 1971.

Roe Type IV Triplane replica BAPC.6

Built 1964, contemporary non-flying replica on original lines.

This aircraft is a faithful replica of A. V. Roe's pioneering aircraft, it is the property of The Aeroplane Collection and has been on loan to the Museum since it opened.

Republic F-84F-46-RE Thunderstreak FU-6

Built 1952, single seat fighter, powered by one Wright J65-W-3 of 7,220 lbs static thrust.

FU-6 started life with the United States Air Force as 52-7133 but was quickly diverted to the Belgian Air Force. It was taken on charge on August 17th 1955 and served faithfully until being struck off charge on December 17th 1970. At this time it had a total of 2,447 hours thirty five minutes flying time. It was stored, along with many other F-84s, at Koksijde until being acquired by the Museum. It was flown into Rochford on November 14th 1973.

SAAB J-29F 29640 '0-8'

Built mid-1950s, single seater fighter, powered by one SFA RM-2B turbojet of 4,750 pounds static thrust. (Licence-built De Havilland Ghost 50).

This 'Tunnen' (Barrel) was flown into Rochford on June 13th 1967 as a gift from Sweden to the British Historic Aircraft Museum. This museum folded and the aircraft was left for a long time out in the open. It was taken on by the HAM and has been refurbished into the colours of F20 Wing of the Royal Swedish Air Force (with whom it once served).

Saunders-Roe Skeeter AOP. Mk.12 XL811

Built 1959, constructor's number S2/5096, aerial observation post, powered by one De Havilland Gipsy Major 215 of 215 horse power.

Completed at Eastleigh from February 6th 1959, XL811 was ready for collection by the Army Air Corps at Middle Wallop twenty one days later. From 1960 to 1963 it served as 'H' of No.651 Squadron. During this time it was to suffer a Category 4 (Beyond repair on site) accident and was under repair (apparently on site despite the allocation) from April 19th 1960 until July 7th 1960. From March 1965 it was with No.652 Squadron and No.17 Flight at Minden in West Germany, followed by the 9/12 Lancers, also in West Germany. It was flown to No.15 Maintenance Unit at Wroughton and was accepted there on December 6th 1967. It was declared a non-effective airframe on May 27th 1968 and put up for disposal, being bought by the Northern Aircraft Preservation Society on December 2nd 1968. It was used by NAPS for display at several northern air displays until being loaned to the Museum in 1972. NAPS are now known as The Aeroplane Collection.

Saunders-Roe P.531 Mk.1 XP165

Built 1960, general purpose light helicopter, powered by one Blackburn Turbomeca Turmo 600 turboshaft of 325 estimated shaft horse power.

This aircraft started out in life as a development machine but was later brought up, as near as possible to Scout standards. It first flew on August 4th 1960 and was used by Blackburn and General Aircraft for trials on its Turbomeca Turmo 600. It then gained its red and white colour scheme and became part of the Empire Test Pilots School, based at Farnborough, and coded '5'. It was retired in 1972 and appeared on the dump at the Royal Aircraft Establishment at Farnborough by January 1973. It was acquired by the Museum in July 1975.

Stampe et Vertongen SV-4B G-AWIW

Built 1947, constructor's number 532, basic trainer, powered by one De Havilland Gipsy Major Mk.10 in-line piston engine of 145 horse power.

Built in Antwerp by Stampe et Renard to a French order powered originally by a 140 horse power Renault in-line, registered F-BDCC. It was imported into the United Kingdom in February 1969 and converted by Rollasons to the powerplant listed above. This aircraft was one of the founding mounts of the 'Rothmans' aerobatic team and served with them until being replaced by Pitts S-2As. It was retained by Rothmans of Pall Mall Limited and displayed for a while at the Colerne Museum. It was flown from Colerne to Rochford on October 25th 1974 and is now displayed at the HAM, in 'Rothmans' colours and on loan from the sponsors of that now defunct team.

Thruxton Gadfly HDW-1 G-AVKE see photo 31

Built 1967, constructor's number HDW.1, experimental autogyro, powered by one Rolls-Royce/ Continental IO-340-A 'flat-four' piston engine of 165 horse power.

This aircraft was designed by Mr. E. Smith and was built by the Gadfly Aircraft Company in Andover. It was tested at Thruxton, but never flown, being found to have stability problems. It was put into store at Thruxton from 1968 until 1971, moving back into Andover until bought by Mr. Elfan ap Rees in November 1972. It was loaned to the HAM from January

1973. It is now part of the British Rotorcraft Museum, but remains on loan to the HAM.

Westland WS-51 Widgeon Series 2 G-APTW

Built 1959, line number WA/H/150, general purpose light helicopter, powered by one Alvis Leonides 521/2 piston engine of 500 horse power.

G-APTW has spent the majority of its life with its builder's, Westland Aircraft. It was with them briefly from birth in June 1959 until being leased to Executaire of Elstree from July for a short period. After this it was with Westlands until 1973 on demonstration, and later general communications work, being replaced eventually by a Gazelle. It was then used by Helicopter Hire of Rochford, who presented the aircraft, on loan, to the Museum in late 1976.

Notes : The following incomplete airframes are on display :-

English Electric Lightning F. Mk.1 XG325 nose section only, on loan from No.1476 Squadron, Air Training Corps, Rayleigh.
Percival Proctor IV G-AOBW sectioned fuselage only, wings donated to NP303 (see above). Ex NP339, delivered by road on March 19th 1967 from Stanmore.
Short S.16 Scion II G-AEZF sections of airframe only, owned by the Anglian Aviation Society. Pobjoy-built in 1937, c/no. PA.1008, flew during War as M-5, withdrawn in May 1954.
Vickers 812 Viscount G-AVHE nose section only. c/no.363, ex Continental Airlines N251V, later Channel Airways. Broken up at Rochford June 1972. Total time 28,142 hours.

The following aircraft are in store :-

De Havilland DH.89A Dragon Rapide G-AIUL c/no. 6837, ex NR749, withdrawn from use in 1970. Acquired by the Museum in 1972, still crated.
De Havilland DH.115 Vampire T. Mk.11 XD527 '46' Acquired from Aviation Traders Limited, Rochford. Built at Christchurch, c/no. 15275.
Hawker Sea Fury FB. Mk.11 WJ244 '026' Dismantled, incomplete. Acquired from Hawker Siddeley from Dunsfold, ex FRU. Possibly to go to the Kent Fighter Museum.
SNCAN SV-4C Stampe Coupe G-AYLK Wreck, c/no. 673, ex F-BDNR, crashed in Eire, circa 1971.
Westland WS-51 Dragonfly HR. Mk.3 WG670 Burnt out in hut fire July 1972, remains.

HUMBERSIDE AVIATION MUSEUM

Situated in the picturesque Gardens of Elsham Hall, the Humberside Aviation Museum is only four miles from Brigg and forms a triangle with the wartime bomber bases of Elsham Wolds and Kirmington (now Humberside Airport), and will be in easy reach of the Humber Bridge when it is completed.

Run on a voluntary basis by members of the Humberside Aircraft Preservation Society this small open air museum was opened towards the end of 1976.

Apart from several aircraft on view, there are also a large number of support exhibits recovered by members or donated by the general public, these include an almost complete Merlin XX engine built in March 1939.

The Museum is open from April to October from 1100 to 1800 hours Saturday and Sunday and on Bank Holidays. Admission is 5p, this includes a write-up on the Society and Museum.

Address : Humberside Aircraft Preservation Society, 66, Emfield Road, Scartho, Grimsby, South Humberside DN33 3BZ.

Auster 5 G-AKWT

Built 1944, constructor's number 998, light cabin monoplane, powered by one Lycoming O-290-3 'flat four' piston engine of 130 horse power, not fitted on exhibit.

Built at Rearsby in 1944 as a Taylorcraft Auster AOP. Mk.5 serial MT360, this aircraft was civilianised, under the designation Auster 5, in 1948, when it was registered as G-AKWT. Only a month after its registration it crashed at Tollerton on August 7th and was written off. Reduced to a bare fuselage and tail units this aircraft was rescued from prolonged store at Stroxton Lodge and moved to Elsham Hall where its restoration is underway.

Auster J/1N Alpha G-AIJI

Built 1946, constructor's number 2307, light cabin monoplane, powered by one De Havilland Gipsy Major 1 in-line piston engine of 120 horse power.

G-AIJI was registered on April 18th 1947 as an Auster J/1 Autocrat with a Blackburn Cirrus Minor 2 of 100 horse power. It was later converted to J/1N Alpha standards, basically the change of engine. The aircraft is known to have visited Malta and West Africa in its travels. Whilst based at the local Kirmington Airport G-AIJI was blown over and wrecked by strong winds on January 125h 1975, and was cancelled from the register on March 12th. It was declared beyond economical repair and was stripped of all usable spares. In this form it was acquired by the Museum and its restoration is well underway.

Miles M.38 Messenger 2A G-AHUI

Built 1946, constructor's number 6335, light cabin monoplane, powered by one Blackburn Cirrus Major 3 in-line piston engine of 155 horse power, not fitted to exhibit.

Like many Messengers, G-AHUI was built at Newtownards in Ulster. It was registered to Derek Crouch (Contractors) Limited at Peterborough on December 12th that year. In February 1954 it gained its last recorded owner, Mr. E. C. Francis. It was withdrawn from use at Wolverhampton in September 1960 and remained in a derelict state there for some time. Until 1975 its history is then obscure. In early 1975 it was acquired by the now defunct Stockport Aircraft Preservation Society and moved into their store at Handforth. By this time the aircraft was only a complete fuselage aft of the engine firewall. In 1976 the project was taken over by the Humberside Aircraft Preservation Society who moved it initially to their store at Goxhill before installing it at Elsham Hall. It is intended to restore the fuselage to representative colours of a wartime Messenger.

Miles M.38 Messenger 2A G-AKBM

Built 1947, constructor's number 6704, other data as above.

This aircraft was discovered lying in a local scrapyard by the Museum, having had the insult of a scrapped car falling across its back. It was rescued and is now under restoration at Elsham Hall. Like its sister above it has been reduced to a fuselage, but has managed to retain its tail surfaces. It too was built at Newtownards and was registered as G-AKBM on August 12th for the Plymouth and District Aero Club, based at Roborough. In 1954 it was placed second in the Siddeley Challenge Cup race, yielding an average speed of 129 m.p.h., in the hands of Flight Lieutenant D'arcy. It was withdrawn from use in May 1961 and languished at Weston-super-Mare until, presumably, being bought by the local scrappy.

Miles M.65 Gemini 7 G-AKHZ

Built 1947, constructor's number 6527 (but see below), light cabin twin, powered by two De Havilland Gipsy Major 10 Mark 2 in-line piston engines of 145 horse power each.

This aircraft is a rebuild using three (and possibly more) aircraft. It was initiated by the Stockport Aircraft Preservation Society at their store in Handforth in 1975, the bulk of the components having been obtained from the hangars at Sywell. The rear fuselage comes from Gemini 1A G-ALUG, constructor's number 6320 and built in 1946 as EI-ACW, returning to the United Kingdom, as G-ALUG, the following year. It was withdrawn from use at Sywell in June 1964 and progressively reduced to spares. The majority of this exhibit comes from 1A G-AKHZ, registered as such on November 7th 1947. In April 1957 it was converted to the powerful Mark 7 variant. It was withdrawn from use at Sywell in January 1964 and also broken up, although much more remained of this aircraft. Some small parts of 3A G-AMME are also known to have been incorporated. As a Mark 7 this aircraft (G-AKHZ that is) was used by the fleet of Shell-Mex and BP and it is hoped to complete the aircraft in their colours.

Westland WS-51 Dragonfly HR. Mk.3 WP503

Built 1953, line number WA/H/88, general purpose naval helicopter, powered by one Alvis Leonides 521 piston engine of 500 horse power.

WP503 flew for the first time on August 6th 1953, from Yeovil. It was taken on charge by the Royal Navy at Gosport on September 4th 1953. It served with the Station Flight at Lossiemouth from the middle of that month until November 1955. It moved to Lee-on-Solent in December 1955 and was placed into storage. From May 1956 it was attached to the aircraft carrier H.M.S. 'Bulwark' until January 1957. It suffered a Category 4 accident and was moved to Donibristle and remained there until 1958 when it was sufficiently healed to be loaned to the Royal Air Force. It was based at North Coates in 1960. It was delivered to the Royal Aircraft Establishment at Thurleigh in 1960 and served there until 1962 on various test duties. It was flown to Stansted on October 12th 1962 and joined the Fire School there for crash rescue training. As such it was not burnt (the fate of most aircraft that end up with the School). It was donated to HAPS and moved to Elsham Hall by road on June 16th 1976.

IMPERIAL WAR MUSEUM

This large building in Lambeth Road (built circa 1815), the former Bethlem Royal Hospital for the Insane or Bedlam, houses the Museum's collections relating to the two World Wars and other military operations involving Britain and the Commonwealth since 1914. A wide range of weapons and equipment is on display : aircraft, armoured fighting vehicles, field guns and small arms, together with models, decorations, uniforms, posters, photographs and paintings.

The collection of aircraft at Lambeth Road is only a small part of the collection now on charge with the I.W.M., this can be found at Duxford airfield near Cambridge and can be found in this work under the heading 'Duxford'.

The Museum is open Monday to Saturday 1000 to 1750 hours, and on Sunday from 1400 to 1750 hours. It is closed Good Friday, Christmas Eve, Christmas Day, Boxing Day and New Year's Day. Admission free.

Underground stations nearest are Lambeth North or Elephant and Castle. The following bus services have stops close by : 3, 12, 53, 59, 159, 10A, 44.

Address : Imperial War Museum, Lambeth Road, London SE1 6HZ.

Telephone : 01-735 8922.

Bristol Type 96 F.2B Fighter E2581 see photos 35 & 36

Built 1918, two seat fighter, powered by one Rolls-Royce Falcon III of 275 horse power.

The British and Colonial Aeroplane Company delivered this aircraft to the Royal Air Force on September 16th, 1918, probably to No.5 Aircraft Acceptance Park, at Bristol. It was certainly with No.5 AAP by November, when it moved on to join No.1 Communications Squadron. It only stayed with the communications unit until the end of the year. Its next charge was with the Headquarters Flight of the South Eastern Area, staying here until April 1919. There followed a spell with No.30 Training Squadron, which could only have been a matter of days. In late April 1919 E2581 joined its final operational unit, No.2 Group Communications Flight with which it served as a general 'hack' until being retired in February 1920. It is thought to have been earmarked for the I.W.M. very soon after this.

De Havilland DH.98 Mosquito T. Mk. III TV959 'AF-V'

Built 1945, operational trainer, two Rolls-Royce Merlin 25s of 1,635 horse power.

Built by De Havilland's at Leavesden, TV959 was first with No.13 Operational Training Unit at Middleton St. George, from July 22nd 1945. Here it was coded 'KQ-G'. In 1947 it moved to No.228 Operational Conversion Unit until November 1954 when it was flown to No.27 Maintenance Unit at Shawbury for temporary storage. By December 1955 it had joined the Home Command Examining Unit, based at White Waltham, the unit being responsible for checking the standards of flight crews within the Command. Its final operational service was made with No.3 Civilian Anti-Aircraft Co-operation Unit at Exeter, where it was coded 'Y'. Here it was was used occasionally to supply gunnery practise, but mostly for conversion training for pilots of the unit. After being replaced by Meteor TT.20s, the Mosquitos of the unit were dispersed and TV959 was earmarked for the I.W.M. Space limitations meant that a wing had to be removed (the starboard one) and this was done at Bicester by No.71 MU prior to installation in the Museum. The wing is in store with the Museum should TV959 ever be displayed whole again. It is displayed with the codings 'AF-V'.

Fairey Swordfish III NF370

Built 1944, three seat naval strike aircraft, one Bristol Pegasus 30 radial piston engine of 750 horse power.

NF370 is a 'Blackfish' - Swordfish built by Blackburn Aircraft Company at Sherburn-in-Elmet, being delivered on April 1st 1944. It was issued to the Royal Air Force on February 13th 1945 and was possibly intended for use with No.119 Squadron. Its service use (if it saw any real use at all) is unknown and it was allocated to Naval Air Station Stretton on December 17th 1946 and was noted still to be in store there by at least April 1950. It became part of the I.W.M.'s collection by at the latest 1960.

Focke Wulf Fw 190A-8/R6 733682

Built circa 1943, single seat fighter, powered by one BMW 801 of 1,700 horse power.

Captured at Fassberg, Germany, 733682 was transferred to the United Kingdom by September 18th 1945. It was allocated the Air Ministry identification number 75. It was displayed at Farnborough in November 1945 in 'Mistel' composite aircraft configuration, mounted on top of Junkers Ju 88A-6 werke number 2492 (Air Min 77). It was subsequently dismantled for storage

37 *'Yellow Peril' BP.111A VT935 at Baginton in 1976 awaiting the opening of the Midland Air Museum*
Stuart Howe

38 *Lincolnshire Aviation Museum's Harvard EZ259 lying in the scrap yard at Sandhurst*
via Les Hunt

39 *Nose drooped, WG774 touches down, showing its stalky landing gear 'Flight' F6644 430*

40 *The Museum of Flight's BAC.221 WG774 overflies Farnborough in September 1966 'Flight'*

41 *Thurleigh line-up June 1957 : BP-111A, FD-2 WG777, Bedstead XJ314 'Flight' 35335 S*

until displayed at the Royal Air Force College museum at Cranwell. After this museum was dispersed the aircraft was transferred to Gaydon and Biggin Hill before handing-over to the I.W.M. at Lambeth on extended loan.

Heinkel He 162A-1 Salamander 120235

Built 1945, single seat fighter, powered by one Jumo 004 turbojet.

120235 was serving with I/J.G. 1 of the Luftwaffe when it was captured at Leck in May 1945. It was brought back to the United Kingdom and given the Air Ministry identification number 63. It was stored at various locations before becoming one of the exhibits in the museum at Cranwell, where it remained until the early 1960s. It was later transferred on loan to to the I.W.M. at Lambeth.

Royal Aircraft Factory BE.2C 2699

Built 1916, two seat light scout, powered by one RAF 1A piston engine of 90 horse power.

2699 was built by Ruston, Proctor andCompany and saw service between 1917 and 1918 with Nos. 50, 51 and 192 Squadrons respectively. It is known to have been used in the defence of London against German air raid attacks. In its second log book 2699 had amassed some 398 hours flying time. It is thought to have been singled out for preservation from an early stage of its post-operational career.

Sopwith 2F.1 Camel N6812

see photos 34 & 35

Built 1918, single seat fighter, one Bentley BR.1 rotary piston engine of 150 horse power.

There are some doubts as to the identity of this aircraft, having at one time worn the totally erroneous serial 'F4043' - see the illustrations. N6812 was built by William Beardmore and Company and was destined to become a famous aircraft. Piloted by Lieutenant S. D. Culley it took off from the Lighter H5 (towed by H.M.S. 'Redoubt') on August 11th 1918 in the southern sector of the North Sea. It intercepted and brought down the Zeppelin L.53 over the North Sea. Such sorties were common place and the aircraft was invariably ditched at the end of the mission. To quote the placard in front of the exhibit N6812 "was recovered almost undamaged after landing in water alongside the Lighter". It is thought that the aircraft was then stored (a salt water dunking would not do it any good) and passed on to the I.W.M. Why it was painted as 'F4043' remains unknown. Details on this aircraft's past much appreciated.

Supermarine Spitfire Mk. I R6915

Built 1940, single seat fighter, one Rolls-Royce Merlin II piston engine.

Built by Vickers-Armstrongs at Southampton, R6915 was first issued to No.6 Maintenance Unit, Little Rissington, on July 11th 1940. It was delivered to No.609 Squadron on the 21st. After assumed operational damage it was delivered to No.1 Civilian Repair Unit on October 14th and thence to No.12 MU at Kirkbride, on January 21st 1941. It then joined No.61 Operational Training Unit on July 6th and from the 22nd was officially on charge with No.43 Group. By the 29th R6915 was in the care of General Aircraft at Hanworth. On October 2nd it had a busy day, being on charge with No.45 MU at Kinloss, and then at No.5 MU at Kemble. By April 2nd 1942 it was back with No.1 CRU and was fit for service and awaiting collection on April 15th. On the 28th it moved to yet another MU, No.6 again, and was issued back to No.61 OTU on June 21st 1942. It was involved in a flying accident on April 20th 1943, but is thought to have been repaired on site. By June 13th it was transferred to No.57 OTU, only to receive more damage in a flying accident on September 21st 1943. It was repaired by December 10th and was issued fifteen days later to No.39 MU at Colerne. In 1944 it came out of the MU for a brief spell (until 24th February 1944, when it returned to No.39 MU) with the Royal Naval Development Unit. From No.39 MU R6915 moved to No.82 MU at Lichfield on May 10th and to No.52 MU, Cardiff, on August 25th. It is presumed to have been struck off charge on June 21st 1947, having been previously handed over to the Imperial War Museum on August 26th 1946.

Note : The following nose sections are on display :-

Avro Lancaster B.I DV372 Served with No.467 Squadron (Royal Australian Air Force) from November 1943 until October 1944, coded 'PO-F'. Completed 49 operational sorties.
Hawker Typhoon Mk.1B Serial unknown, came from RAF Halton.
Mitsubishi Model 52 A6/M5 'Zero Sen' Captured in South East Asia in 1945 and used for a while by the Allied Technical Air Intelligence Unit - South East Asia, whose markings it still carries. Was marked with the 'serial' BI-12'.

IRISH AVIATION MUSEUM

In July 1970 Captain Kelly-Rogers, retired Deputy General Manager of Aer Lingus, accepted the invitation of that Company to try and set up an Irish Aviation Museum. Whilst he was an Aer Lingus executive he had often demonstrated this interest in aviation history by his actions and not the least meritorious of these was his responsibility for the design and erection of the Alcock and Brown memorial at Clifden, County Galway.

In six years a considerable amount of Irish aviation history has been garnered in various shapes and forms. Donors of material have been many and generous but donations have largely taken the form of documents, prints, photographs and the like. Hardware, in the shape of aircraft, has been more difficult to come by. Difficult also to exhibit because, as yet, the Irish Aviation Museum has been unable to find a permanent home which it can afford to occupy.

As an interim measure a substantial amount of material so far collected forms the Aeronautical Exhibition at Dublin Airport. It is housed in the Exhibition Centre located in the departures concourse and there is great variety among the exhibits on display. These all have an Irish connection. In addition to much pictorial history there are historical engines, models, instruments and aircraft parts such as a pair of Bristol Fighter wing frames made in the town of Carlow during World War One. The State Services - Air Traffic Control Meteorology and Radio - have their own displays.

Two aircraft are displayed at the Exhibition - a Dragon biplane and the nose of a Viscount 800 series aircraft.

The Aeronautical Exhibition is open daily from 1000 to 1700 hours, except Saturdays and Sundays when it is open until 2000 hours. Admission is 20p for adults and 10p for children under 16 - special rates are available for groups. During the first year of operation which ended in September 1976, some 57,000 persons paid for admission.

Address : Irish Aviation Museum, Hon. Curator, J. C. Kelly-Rogers, Spindrift, Shore Rd., Portmarnock, County Dublin.

De Havilland DH.84 Dragon 2 'EI-ABI'

Built 1936, constructor's number 6105, small feederliner/air taxi aircraft, powered by two De Havilland Gipsy Major in-line engines of 130 horse power each.

The aircraft on display is painted as Aer Lingus's first owned aircraft, EI-ABI 'Iolar', but is in fact a later sister, EI-AFK. EI-ABI was built as G-ACPY for Olley Air Services of Croydon. In February 1935 it passed to Blackpool and West Coast Air Services Limited and then to Railway Air Services Limited. RAS sold it to Aer Lingus and it was delivered to Dublin, as EI-ABI, on May 26th 1936. The following day it undertook the airline's first service, carrying the name 'Iolar' on the nose. In 1938 EI-ABI was sold in the United Kingdom and restored as G-ACPY, it was shot down by enemy aircraft on June 3rd 1941 on a trip between the Scilly Isles and Lands End whilst with Great Western and Scottish Airlines. The aircraft that now portrays EI-ABI, EI-AFK, has never served with Aer Lingus. It was built in 1936 for Air Cruises Ltd., as G-AECZ, passing to Ramsgate Airport Limited (in November 1937), Southern Airways Limited (March 1939) until being impressed into military service as AV982 in May 1940. It returned to civilian life with Air Taxis Limited at Croydon in June 1946. It was sold to J. Cleary and delivered to Eire on March 1st 1950, becoming EI-AFK. It was later registered to Weston Limited. It was cancelled from the Irish register in March 1966. It was overhauled and painted as EI-ABI and flown to Dublin for acceptance by Aer Lingus on September 1st 1967. It was kept in the roof of the Aer Lingus hangar at Dublin until being refurbished for diplay with the IAM by Aer Lingus, also at Dublin.

<u>Notes</u> : The following nose section is also on display :-

<u>Vickers V.803 Viscount EI-AOH</u> constructor's number 180, previously with KLM as PH-VII. Delivered to Aer Lingus on February 25th 1967, withdrawn from use October 31st 1970 after replacement by BAC 111s and Boeing 737s, broken up, nose donated to the IAM.

The IAM also owns the Ferguson Monoplane replica displayed at the Ulster Folk and Transport Museum, which see.

The IAM has very close connections with the Irish Air Corps, and it is hoped to preserve some IAC types. Amongst these is the following :-

<u>Miles M.14A Magister 34</u> constructor's number 1028, ex Royal Air Force N5392, delivered

to the Irish Air Corps on February 22nd 1939 and served until being relegated to instructional aiframe status as Baldonnel on March 11th 1952. It has been completely restored with the help of Aer Lingus.

LASHENDON AIR WARFARE MUSEUM

Situated at Headcorn Airfield, in Kent, the Lashendon Air Warfare Museum was founded in the early 1970s to consolidate and display the considerable aviation history of Kent and the surrounding countryside. The Museum occupies some converted farm buildings at the airfield and, apart from the V.1 and the Vampire, specialises in displaying World War Two exhibits stemming basically from 'aerowreckology' work on the many crash sites of the neighbourhood.

The Museum is open to the public every Sunday from April to October and admission is free. Readers intending to visit are advised to use the correctly marked entrance to Headcorn airfield for the Museum, and not the entrance to the operational airfield. Admission is free.

Address : Lashendon Air Warfare Museum, Headcorn Airfield, Headcorn, Kent, TN27 9HX.

De Havilland DH.115 Vampire T. Mk.11 WZ589

Built 1953, constructor's number 15164, advanced trainer, one De Havilland Goblin 35.

WZ589 was built at Hawarden in 1953 and its service history is little known, apart from serving No.56 Squadron as a 'hack' until being bought back from the RAF by Hawker Siddeley on October 30th 1967. It was put into store at Hawarden, but moved to Woodford in 1971/72. At one time it was earmarked for possible refurbishment and resale to the Chilean Air Force, but this did not ensue. In 1973 it was offered up (along with many other Vampires) for sale to preservation societies and the like. WZ589 was bought by the Kent Branch of the Royal Air Force Association and delivered by road to Headcorn on July 3rd 1973, where it was displayed on loan to the Museum.

Fieseler Fi.103R-IV BAPC.91

Built circa 1942, thought to be an experimental piloted version of a flying bomb, powered by one Argus 014 pulse jet.

This example of the famous V.1 'Doodlebug' is believed to have been a piloted version, and the exhibit shows the cockpit canopy to this day. It is possible that it is a careful adaption of a normal flying bomb version however, but if this is the case then it was done prior to inclusion in the Museum. It is thought to be the machine previously kept at Fort Clarence, Rochester, and the Joint Services Bomb Unit at Horsham. It was refurbished in 1975.

LEICESTERSHIRE MUSEUM OF TECHNOLOGY

This branch of Leicestershire Museums was formed in the later 1960s to record the industrial and technological history of the East Midlands, but following local government reorganisation and financial cut-backs, now restrict coverage to the County of Leicestershire.

Aeronautical items in the collections represent the early history of jet engines and aircraft manufacturing by Reid and Sigrist, Brush, and Taylorcraft/Auster/Beagle. It is hoped the Museum will become the main centre recording the history of 'Auster' designs.

Exhibition space is at present restricted and the specimens on show are continually changing. However displays dealing with the above are planned and will be started in 1977 if finance allows.

The Museum is open 1000 to 1730 hours Monday to Saturday and 1400 to 1730 hours on Sundays. Admission is free except for occasional special events, e.g. Beam Engine or Excavator in steam, when a small charge is made to cover fuel.

There is a publications counter selling Museum and other relevant material.

Address : Leicestershire Museum of Technology, Corporation Road, off Abbey Lane, Leicester.

Telephone : Leicester 61330.

Auster B5 AOP. Mk.9 XP280

Built 1960/61, constructor's number AUS/B5/166 (subject to confirmation), aerial observation post, powered by one Blackburn Bombadier 203 in-line piston engine of 173 horse power.

XP280, a late production aircraft, was issued to No.651 Squadron, Army Air Corps, Middle Wallop in June 1961. This later became No.651 Light Aircraft Squadron and XP280 was detached to the 2nd Royal Tank Regiment Air Troop of the Unit. Flying into overhead cables necessitated a new starboard wing in December 1963. By November 1964 the aircraft was serving with the 1st The Queen's Dragoon Guards Air Troop at Aldergrove and continued so but for a short period with Headquarters Royal Scots 2nd Wing (U.K.) in June 1965. In mid-1966, when the Army were favouring helicopters rather than fixed-wing aircraft, XP280 was transferred to store at No.19 Maintenance Unit at St. Athan where it remained until purchased by the Museum in November 1974. It is in store at the Museum, but it is hoped to display it during 1977.

Note : The following aircraft is maintained in airworthy condition by the Museum at Leicester East Airport.

Taylorcraft Auster 5 Series J/1 Autocrat G-AGOH

Built 1945, constructor's number 1442, light aircraft, powered by one Blackburn Cirrus Minor II in-line piston engine of 100 horse power.

First flown on May 25th 1945 G-AGOH was an Auster Mk.5 airframe with a Cirrus Minor fitted instead of the usual 130 h.p. Lycoming to test this more economical installation before production of a three seat 'Auster' for post-War use. A full prototype with civil cockpit layout and reduced glazing was also made from the material of the pre-War Taylorcraft Plus D prototype G-AFWN (this aircraft is now under restoration in Denmark). From 1945 until the early 1950s G-AGOH was used for continuous Cirrus Minor development flying by Blackburn at Brough. It was then sold and flew with three successive private owners from Leicester East airfield until purchased by the Museum in 1969. It now flies air observation and photography missions for various sections of Leicestershire Museums as well as attending various historic aircraft rallies.

LINCOLNSHIRE AVIATION MUSEUM

The Museum is situated on the A153 road west of Tattershall and Coningsby villages. It is open from 1400 to 1730 hours every Sunday and Bank Holiday from the beginning of May until September. Precise location is the Old Station Yard, Tattershall.

The Museum is run on a voluntary basis by members of the Lincolnshire Aviation Society which holds branch meetings in Boston, Lincoln and Grimsby. It is a member of the British Aircraft Preservation Council.

Containing a wide selection of aircraft, the Museum also has a large supporting display. This comprises mainly of articles that have been found in the surrounding countryside and in The Wash on some of the 'digs' in which the Society has participated, and also from articles that people have turned up with at the Museum. Larger exhibits include a searchlight donated by Butlins at Skegness and various engines. There is also a comprehensive model collection.

The aircraft themselves are in various stages of restoration, the Auster and the Dragonfly currently being worked on. The Dragonfly has regained its rear fuselage and most of its perspex. The Dove, although painted on the top sides, is still in primer underneath and in need of engines.

The reader is warned that the opening of the 1977 season is subject to the continuing of the lease on the Station Yard and that before embarking on a visit should write for current details, enclosing a stamped addressed envelope, to the address given below.

Address : The Curator, Mr. M. Hodgson, Lincolnshire Aviation Museum, Lancaster Farm, Tumby Woodside, Mareham-le-Fen, Boston.

Auster B5 AOP. Mk.9 XK417/G-AVXY

Built 1955, constructor's number AUS/10/92, aerial observation post, powered by one Blackburn Bombadier 203 in-line piston engine of 173 horse power.

XK417 was delivered to No.19 Maintenance Unit at St. Athan on September 18th 1956 before issue to the Army Air Corps Centre at Middle Wallop on May 28th 1957. It went to West Germany and joined No.18 Flight at Detmold from June 9th 1959, transferring to No.9 Flight on August 1st 1961. Exactly a year it joined No.1 Reconnaissance Flight/No.652 Squadron. It returned to No.19 MU on October 26th 1964 for disposal. (Interestingly enough it carried a No.1 (BR) Corps crest on its cowl - was it ever with this unit?) It was declared a non-effective airframe on February 3rd 1966 and had its engine inhibited on June 15th 1967. On November 11th it was registered as G-AVXY to Mr. E. B. E. Pearce, but was never to be converted to civil status. It was noted dismantled at Henstridge in February 1970 and later appeared in Else's scrap-yard in Wisbech. It was acquired from here by the Museum.

De Havilland DH.104 Dove Mk. 1B G-AHRI

Built 1946, constructor's number 04008, air taxi/executive aircraft, powered by two De Havilland Gipsy Queen 70 in-line piston engines of 330 horse power each, not fitted.

G-AHRI was first registered to the Iraq Petroleum Transport Company Limited and was officially handed over to them, as a Mk.1 on the last day of 1946. After service in Iraq it was bought by T. D. Keegan and registered to him at Southend on June 11th 1957 and up-dated to a Mk.1B. On completion of this work it was delivered to Israel and Avitours Air Services Limited, becoming 4X-ARI. It was re-imported in October 1965 by Brooklands Aviation Limited and taken to Little Staughton. It was not made airworthy and remained in store there until it was donated to the Museum, some time after 1971. It was collected and taken to Tattershall on February 3rd 1973.

De Havilland DH.115 Vampire T. Mk.11 WZ549

Built 1953, constructor's number 15141, advanced trainer, one De Havilland Goblin 35.

WZ549 was built at Christchurch and was ready for collection by the RAF on April 10th 1953. On April 4th it was given on free loan to the Controller (Aircraft) and was at No.27 Maintenance Unit at Shawbury by December 27th 1953. It was taken to Marshalls at Teversham on May 7th 1954 for modifications, including the fitting of a clear-view canopy, this work was finished on January 25th 1955 and WZ549 was back at No.27 MU two days later. It was moved to No.22 MU at Silloth on February 8th 1955, ready for issue to its first unit, the Ferry Training Unit at Benson from May 6th. This lasted until January 2nd 1958 when it was issued to No.10 MU at Hullavington, moving to No.5 MU at Kemble on December 9th. On January 27th 1959 it was on charge with No.8 Flying Training Schoolcoded '55' and remained so until 1961. It was briefly with De Havillands for a re-finish between October 2nd and November 16th 1959. On July 4th 1961 it re-acquainted itself with No.27 MU, until re-joining No.8 FTS, still as '55', from April 23rd 1963. From March 11th 1964 it was flying with No.1 FTS as '68'. It went back to Shawbury on November 4th 1964, but this time for service with the Central Air Traffic Control School. It was damaged in a flying accident (Category 3) on November 20th 1964 was was presumably repaired on site by No.27 MU from that date until January 18th 1965, on which date it re-joined the CATCS, now with the code 'F'. It was flown to Coningsby on December 19th 1970 for ground instructional use, for which purpose it was allocated 8118M. It was acquired by the Museum shortly afterwards.

De Havilland DH.115 Vampire T. Mk.11 XD447 '33'

Built 1954, constructor's number 15384, advanced trainer, one De Havilland Goblin 35.

XD447 was built by Fairey's at Ringway and was handed over to the RAF on November 26th 1954. It was delivered to No.48 Maintenance Unit at Hawarden on November 3rd. It was issued to No.5 Flying Training School at Oakington on May 4th 1955, until moving to No.10 MU at Hullavington on August 11th 1958. Fifteen days later it joined the Royal Air Force College at Cranwell, as '50', serving there until January 24th 1962 when it joined No.8 FTS at Swinderby, as '33'. It was retired to No.19 MU at St. Athan on August 29th 1963 and stored there until December 1st 1967 when it was bought by Hawker Siddeley and moved to Hawarden for storage. It was moved in 1971/72 to Woodford and eventually offered to the British Aircraft Preservation Council. It was acquired by the Museum and arrived by road at Tattershall on May 25th 1973.

Mignet HM.14 Pou du Ciel BAPC.43

Built 1930s, single seat ultra light, powered by one Scott A.2S 'Flying Squirrel' two cylinder piston engine of 25 horse power.

This aircraft was built by Mr. A. W. Troop of Wellingore in the 1930s and is known to have 'hopped' under the power of its engine. It was never officially registered and very soon put into store at the Troop household in Wellingore. It remained in store until 1975 when it was presented to the Museum by the builder's family. It is displayed inside the engine shed at Tattershall and serves as good comparison to the Gnome (see below).

Slingsby T.8 Tutor BGA.794

Probably built 1957 or 1958, two seat glider.

Very little is known of this example, although it is believed to have been built by its owner, J. Whitelam, from spare parts or a Slingsby kit. The British Gliding Association number indicates its origin as being in 1957 or 1958.

Ward P.45 Gnome G-AXEI

Built 1966, constructor's number P.45, single seat ultra-light, powered by one converted

Douglas motor-cycle engine.

This aircraft was designed and built by Mick Ward at North Scarle, Lincolnshire, in 1966. It first flew at Wigsley on August 4th 1967, unregistered. It was the United Kingdom's smallest piloted aircraft. It was sold to R. A. Fixter in 1969, and registered as G-AXEI. Mr. Fixter presented the aircraft to the Museum after a brief flying career.

Westland WS-51 Dragonfly HR. Mk.3 WH991

Built 1953, constructor's number WA/H/67, general purpose naval helicopter, powered by one Alvis Leonides 521 piston engine of 500 horse power.

WH991 first took to the air on February 4th 1953, at Yeovil. It was taken on charge by the Fleet Air Arm at Gosport twenty days later. It joined the H.M.S. 'Illustrious' Ship's Flight April 20th, but was believed to have been damaged on May 1st and moved by road to Westlands for repairs. It flew again (from either Yeovil or Yeovilton) on October 27th. It was delivered to the Aircraft Holding Unit at Gosport on November 25th, joining No.705 Squadron there on January 14th 1954, wearing the coding '708/GJ'. It arrived at Donibristle on April 29th 1955 in preparation for joining the Ship's Flight of H.M.S. 'Centaur' on December 18th. It was at Lee-on-Solent from May 16th 1956 for a series of modifications before being issued to the Station Flight at Eglinton on June 14th, and gaining the code '917/GN'. It is then reported to have been converted into a HR. Mk.5, but this cannot be confirmed. Whatever work was done on it was finished on May 1st 1958. On March 9th it was transferred (almost certainly by ship) to Trinidad for H.M.S. 'Vidal', it was back in the United Kingdom by July 1960. It joined No.700 Squadron at Yeovilton on May 2nd 1961 and was involved in trials with the Indian Navy aircraft carrier 'Vikrant', before joining No. 705 Squadron again, this time at Culdrose, probably coded '548'. It is reported to have been the last Dragonfly to serve with No.705 Squadron and may have ended up with the Culdrose Station Flight, gaining the coding '914/CU'. It was in store at Culdrose in 1964, moving to Fleetlands in 1967. It was struck off charge as being sold to W. Sparke of Taunton on June 8th 1970. It was sold in Cyprus but was damaged at London Docks and was bought by Else's scrap-yard at Wisbech. It is now on loan to the Museum from Roger Windley.

Notes : The following incomplete airframes are displayed :-

English Electric Canberra B. Mk.2 WD954 nose section only, ex No.71 Maintenance Unit, Bicester travelling display.
Mignet HM.14 Pou du Ciel fuselage only, donated in 1975, from Sleaford, damaged in fire.
North American Harvard III EZ259 cockpit section only. Built in 1942 at Dallas as 42-84182. Taken on charge at Wingfield (South Africa) 23/ 2/45. In store at Lossiemouth by 1/49 and then at Arbroath, to Skylines yard at Sandhurst in 1956. Reported to have been coded '347' but '247' much more likely. Bought by the Museum via NAPS and restored. *see photo 38.*
Percival Proctor IV G-AOBI remains only in use in the rebuild of NP294 (see below). Ex NP358, derelict with Air Scouts at Halfpenny Green. Collected June 1974.

The following aircraft are owned by the Museum but are kept elsewhere :-

Miles M.65 Gemini 1A G-AKER fuselage and wings, c/no.6491, kept at Goxhill.
Percival Proctor IV NP294 'TSMJ' fuselage only, see G-AOBI above, under restoration at Friskney.

MIDLAND AIR MUSEUM

It was inevitable that sooner or later the Midlands, which has seen so much of the development of aviation in Britain over the past seventy years, would have a Museum of its own; a Museum in which those achievements could be recorded and tribute paid to the men and women who made them possible.

However the establishment of such a Museum has not been achieved overnight. It has taken ten long years of determined effort on the part of amateur enthusiasts, giving their time to building up the collection, patiently restoring exhibits which, it was hoped, would one day be displayed to the public. But now, the results of those ten hard years of work can be seen in the Midland Air Museum at Coventry Airport, Baginton.

Based on an operational airfield, only a few yards from a busy main runway, the Museum aims not merely to display historic aircraft, but to do so in such a way that the visitor is able to imagine those aircraft in their original setting. Ancilliary equipment such as airfield radar, motor vehicles and so on, help to recreate scenes in which aircraft are seen in context rather than in isolation.

Nearly thirty aircraft are available for eventual exhibition. Some of these are in displayable condition and will form the initial exhibits whilst others require considerable restoration and must await their turn.

Coventry Airport is well sign-posted and the Museum is situated off Rowley Road, between Baginton village and the A45, opening each Sunday between April and October and on other days to be advertised. For visitors arriving by air, Coventry Airport is open 0730 to 2200 hours, and the radio frequency is 119·25.

For further information and/or details of membership please contact the address below.

Address : The Secretary, Midland Air Museum, 70, Broomfield Road, Earlsdon, Coventry CV5 6JX.

Telephone : Coventry 713258.

Boulton Paul BP-111A VT935

see photos 37 & 41

Built 1950, delta wing research aircraft, powered by one Rolls-Royce Nene turbojet of 5,100 pounds static thrust.

VT935 was built in Wolverhampton to Specification E27/46 calling for a delta-wing in-flight research vehicle. The 'Yellow Peril' first flew on October 6th 1950 with a 45 degree wing. It appeared at the 1951 Society of British Aircraft Constructor's Display at Farnborough and also at that year's Paris Air Salon at Le Bourget. In 1953 it was modified, mostly internally, from BP-111 standard to BP-111A standard, and reflown in the new guise on July 2nd. It was used by the Aerodynamics Flight of the Royal Aircraft Establishment at Thurleigh until it was withdrawn from use on June 20th 1958. Up to that time it had conducted a considerable number of trials and had flown at speeds up to Mach 0·93. In April 1959 it was bought by the College of Aeronautics, Cranfield, and used by them as an instructional airframe and a design sample. By 1974 the College, by then termed the Cranfield Institute of Technology was experiencing many calls on the space that was occupied by its historic airframes, and the BP-111A was amongst those disposed of. It was bought by the Midland Aircraft Preservation Society (founders of the MAM) who at first entered into negotiations with Boulton Paul to display it outside their plant at Wolverhampton. However, this plan came to nought, and by the time this was realised the seed of the Museum at Baginton had been planted. VT935 was moved by road to Baginton on July 13th 1975 to form the nucleus of the present day Museum.

De Havilland DH.100 Vampire F. Mk.1 VF301

Built 1946, single seat fighter, powered by one De Havilland Goblin 2 of 3,100 lbs. s. t.

VF301 was built by English Electric and was delivered to No.226 Operational Conversion Unit on September 27th 1946 for a brief spell before entering No.33 Maintenance Unit at Lyneham on October 30th. It then was issued to No.595 Squadron from February 19th 1947, followed by No.631 Squadron from September 24th 1948. It returned to No.33 MU on June 19th 1950. It was returned to De Havillands on February 9th 1951 for a refurbish, this work being completed when it was delivered to No.5 MU at Kemble on May 21st. On August 29th it became a part of No.103 Flying Refreshrer School, until going to No.19 MU at St. Athan on November 21st. It reappeared on July 21st 1952 and joined No.208 Advanced Flying School only to receive a serious Category 3 (Repair beyond unit capacity) flying accident on October 28th. Repairs were begun on site by No.49 MU from Colerne from the following day, but VF301 had to be removed to No.5 MU at Kemble on January 12th 1953 for repairs to continue, it moved on to De Havilland's on the 26th. The worked was completed on April 30th, only to have the aircraft declared a non effective airframe on May 13th!! It was moved to Debden and was taken on charge there on August 25th 1953, becoming 7060M. It was displayed on the gate there until acquired by MAPS in 1973 for £100.

Fairey Ultra Light Helicopter G-APJJ

Built 1958, constructor's number F.9428, two seat ultra light helicopter, powered by one Blackburn-built Turbomeca Palouste BnPe 2 turbo-compressor.

The Fairey Ultra Light Helicopter was built to Specification H.144T to give the Army a small helicopter for general light duties. Eventually the Ministry of Supply dropped its backing of the project and the Fairey company were left to continue on their own. G-APJJ was built at Hayes and flew for the first time at White Waltham in 1958. It was the fifth flying aircraft. After testingwas completed by Fairey, and the design was not put into production, G-APJJ moved on the the Royal Aircraft Establishment at Thurleigh and used for non-flying experiments on rotor behaviour. One of these tests involved mounting G-APJJ on top of a double decker bus and studying its rotors whilst the bus was in motion. It later became the property of the Cranfield Institute of Technology. Like the BP-111A above, it was squeezed out by the CIT's expansion and was bought by MAPS, who used its portability to good effect by taking it along with their sales stall to several Midland events.

Mignet HM.14 Pou du Ciel G-AEGV

Built 1936, constructor's number EMAC.1, single seat ultra light, powered by one Douglas twin cylinder piston engine of 750 c.c.

G-AEGV was built by the East Midlands Aviation Company at Sywell in 1936. It is thought to have flown, but it is not known how well. It was cancelled from the Civil Register in December 1937. It was put into store and parts of it survived beyond World War Two. These components were rescued by MAPS and the airframe was subjected to a major rebuild. This was completed by the late 1960s and the 'Flea' has since been used as a mobile attraction at the Society display stand.

Vickers 668 Varsity T. Mk.1 WL626

Built 1956, multi-purpose aircrew trainer, powered by two Bristol Hercules 264 radial piston engines of 1,950 horse power each.

WL626 was built at Hurn, and first flew from there on April 15th 1953. It was delivered to No.201 Advanced Flying School fourteen days later, at Swinderby. Here it was coded 'P'. Like most Varsities, it went to soldier through many years of training of one kind or another without any major accident. WL626 served with : No.2 Air Navigation School, coded 'Z', then with No.1 ANS, coded 'O' and later 'U' based at Topcliffe and later Stradishall and finally with No.6 Flying Training School at Finningley. It was retired from service with No.6 FTS and bought by Graham Vale on behalf of the Museum and flown into Baginton on April 12th 1976.

Notes : Along with the above aircraft the following external replica aircraft are displayed, Messerschmitt Me 109 (Ha 1112) BAPC.67, Hawker Hurricanes BAPC.68 and BAPC.72 and Supermarine Spitfire BAPC.69.

The following aircraft are in the Midland Aircraft Preservation Society's collection and may be exhibited at the Museum in the future :-

Crossley Tom Thumb BAPC.32 High wing cabin monoplane, built in Banbury in 1937, not completed. In store in Coventry.
De Havilland DH.83 Fox Moth G-ACCB c/no. 4042, built 1933, ditched off Southport 25/ 9/56. Used for spares for G-ACEJ and G-AOJH. Fuselage only, in store in Stratford-upon-Avon.
De Havilland DH.89 Dragon Rapide G-AJBJ c/no. 6765, built as Dominie I NF844, dismantled and stored at Squires Gate from September 1961. Fuselage only, in store at Baginton.
Flettner Fl 282 Kolibri 28368 German single seat helicopter, twin rotor, captured in 1945, later to College of Aeronautics at Cranfield. Frame only, in store in Coventry. Possibly to return to Germany for display in Munich.
Folland Fo.141 Gnat F. Mk.1 XK741 Single seat fighter prototype, built 1955, used for testing at Dunsfold and later at Wolverhampton. Being refurbished at Leamington Spa.
Grunau Baby II VT921 Secondary glider, being refurbished at Honington.
Hawker Sea Hawk FB. Mk.3 WM994 A2503 previously in use with the Cranfield Institute of Technology. In Swansea being refurbished by Mr. H. Webborn.
Humber-Bleriot XI static replica BAPC.9 Built about 1959, ex Shuttleworth Trust and display at the Fleet Air Arm Museum, Yeovilton. In store in Coventry.
Kranich II BGA.964 Glider, ex SE-STF, in store in Stratford-upon-Avon.
Luton LA.4 Minor G-BAPC c/no. MAPS-2, wings only for possible rebuild to flying condition. Possibly ex G-AEPD, c/no. LA.3, stored in Coventry.
Mignet HM.14 Pou du Ciel BAPC.27 Largely contemporary build, by M. J. Abbey in Coventry, In store in Stratford-upon-Avon.
Mignet HM.14 Pou du Ciel 'G-AEOH' Built by R. C. Streather in Sutton Coldfield, c/no. RCS.1. One Scott A.2S, flown 1937. Rebuilt (including new fuselage). In store in Coventry.
Miles M.38 Messenger 4A VP-KJL ex G-ALAR and RH371, from the Shuttleworth Trust, fuselage only, In store in Coventry.
Nyborg TGN.III BAPC.25 Single seat glider designed and built by T. G. Nyborg of Worcester. In store in Stratford-upon-Avon.
Parnall Pixie III G-EBJG 1926 two seat light aircraft, conversion of biplane of 1924, incomplete, in store in Coventry.
Percival Q.6 Petrel G-AFFD In store at Duxford, which see.
Slingsby T.7 Cadet BGA.804 Training glider, in store in Coventry.
Wheeler Slymph G-ABOI Built 1931 by A. H. Wheeler, not flown, c/no. AHW.1. On loan.

R. J. MITCHELL HALL

There have been plans to build a memorial museum to R. J. Mitchell in Southampton for many years but rising costs caused the original scheme to be abandoned and it became obvious that if anything was to be done then a less ambitious project would have to be undertaken.

In September 1975 a new committee was formed with the backing of No.424 (Southampton) Squadron Air Training Corps and an old dance hall was leased from the City Council for conversion to a museum, this involved extensive internal alterations and the building of an extension to house the shop and offices. Most of this work was done by volunteers from the ATC Squadron using materials donated by local companies. In February 1976 most of the major constructionwork was completed and the aircraft were delivered and erected by No.71 Maintenance Unit in time for the 40th Anniversary of the Spitfire in March when the Museum was opened for two days. It then closed to enable the rest of the exhibits, consisting of documents, photographs and models illustrating all of Mitchell's designs from 1916 to 1937 to be assembled and displayed. On June 11th 1976 Dr. K. Gordon Mitchell B.Sc. Ph. D. (the son of R. J. Mitchell) officially opened the Museum.

The Museum is situated in the centre of Southampton opposite the Civic Centre. Access is from a footpath (Kingsbridge Lane) leading to Central Station. There is plenty of public car parking close by.

The Museum is not open Mondays. Opening times Tuesday to Saturday are 1000 to 1630 hours and Sunday 1400 to 1700. Admission is adults 20p and children 10p.

Address : The R. J. Mitchell Hall, Kingsbridge Lane, Southampton SO1 OGB.

Supermarine S.6A N248

Built 1929, high performance racing float-plane, powered by one Rolls-Royce 'R' piston engine

This aircraft was built by Supermarine's at Woolston in 1929 as an S.6 for entry into the 11th Schneider Trophy Race, to be held at Calshot that September. During the actual race, the aircraft, race number 8, was disqualified as its pilot, Flying Officer R. L. R. Atcherley had cut inside one of the pylon markers. Even so he managed to set up a new World Speed Record for both the 50 and 100 kilometer closed circuit of 331 and 332 m.p.h. respectively. In 1931 N248 underwent detail modifications, including rudder balance and increased bracing of the rear fuselage, thus becoming an S.6A. In June 1931, during a practice flight part of the engine cowling became dislodged in flight and Flight Lieutenant Hope was forced to make an emergency landing in the wake of a liner, which forced the S.6A to overturn. Damage was not as bad as at first thought and the floatplane was ready for the 12th Schneider Trophy Race, which was again held at Calshot. This time N248 was only the reserve aircraft, and carried the racing number 4. After this N248 was stored, until 1942 when it was unearthed for use in the film 'The First of the Few', about the life of Mitchell, N248's designer. For the film N248 was dressed as S1596, an S.6B. After the film N248 was again stored, still in the guise of S1596, and the craft's original identity was beginning to slip into history, its true self only coming to light after renewed interest with the opening of the Mitchell Hall in 1976. It 1946 it was brought out of mothballs and refurbished (as S1596) for presentation to Southampton Corporation. It was displayed for several years on the Royal Pier in Southampton (with sorties to London most Septembers for Battle of Britain Week, and to take part in the 1951 Festival of Britain). In 1966 it left the Pier for the last time and was taken into the custody of the RAF once again, and stored at Henlow. It was displayed briefly at the Anniversary of the Aeroplane and Armament Experimental Establishment at Boscombe Down in 1970. After refurbishing S1596 (as it was still pretending to be) was brought by road to Southampton on February 7th 1976 and installed in the R. J. Mitchell Hall. This aircraft has now been repainted in its original colours, as N248.

Supermarine Spitfire F. Mk.24 PK683

Built 1946, single seat fighter, powered by one Rolls-Royce Griffon 85 of 2,050 h.p.

PK683 was built at Castle Bromwich in 1946 and was first issued to No.33 Maintenance Unit at Lyneham on August 13th. By June 30th 1950 PK683 had been transferred to No.47 MU Sealand and crated for shipment. On August 10th the Spitfire arrived at Birkenhead docks and was despatched six days later to the Far East Air Force, arriving in Singapore on September 13th. The 30th Maintenance Base Far East was preparing the aircraft for use again after its sea voyage. On February 28th 1951 it was back at the Maintenance Base for repairs of an unknown nature, sustained whilst serving with a unit also unknown. On July 31st PK683 was transferred to the Singapore Auxiliary Air Force, only to suffer a Category 3 (Repair beyond unit capacity) accident on February 29th 1952. Almost a year later, on February 10th 1953 PK683 was taken on by No.390 MU at Seletar and was struck off charge on April 15th 1954. It was allocated to instructional airframe use, as 7150M, and was issued to the Malayan Air Training Corps, latermoving on to become the gate guard at Changi, where it received the codes 'QV-A'. In 1970 it was dismantled and flown back to the United Kingdom in April on board a Belfast C. Mk.1 of No.53 Squadron. On April 16th No.71 MU at Bicester took charge of it, the aircraft having been singled out for the Southampton Mitchell project. The problems faced by this project meant that the RAF had to look after it until such time as it could be taken over at Southampton, so PK683 entered another period of storage. On June 30th 1972 it was moved to Kemble, but had passed on to Colerne by that November. With the untimely demise of the Colerne collection the now much-travelled Spitfire was removed by road to Shawbury on August 21st 1975. By January 19th 1976 PK683 was moved again, this time to Kemble for preparation for presentation to the R. J. Mitchell Hall, where it arrived on February 7th and has been on display ever since.

R. J. MITCHELL MEMORIAL

This project was the first memorial to the immortal designer of the Spitfire, Reginald J. Mitchell. He was a native of Stoke-on-Trent and as such many people from the town felt that his work and life should be marked in some appropriate way. To this end the town of Stoke has erected a memorial so unique that it really merits an inclusion in this work, as it is a museum as much as it is a memorial. The centre piece for this memorial is, of course, the Spitfire, which is mounted on a plinth. To protect it from the outside world (in all its forms) a clear 'greenhouse' has been erected around it. This structure is used to house displays of Mitchell's life, and the products of his fertile mind. The Spitfire was moved on site on February 15th 1972 and the whole memorial opened in an official presentation on June 28th that year.

The R. J. Mitchell Memorial can be found in the Hanley district of Stoke, in Bethesda Road, which is close to the main Police Station. It is unmanned, and as such never closes.

Supermarine Spitfire LF. Mk.XVIe RW388 'U4-U'

Built 1945, single seat fighter, powered by one Rolls-Royce Merlin piston engine.

This aircraft was first issued to No.6 Maintenance Unit at Brize Norton on July 18th 1945 and went on from preparation there to join No.667 Squadron on August 2nd. It was given the code 'U4-U' at this time, and it wears these again today. By November 30th RW388 had suffered a flying accident and was moved to No.411 Repair and Salvage Unit's charge on December 12th. After repair it was held by No.29 MU at High Ercall, from May 9th 1946, moving back to No.6 MU on January 12th 1949. By March, the 23rd, the aircraft was on charge with the Handling Squadron, for testing, only to return to No.6 MU on April 29th. On June 2nd RW388 became part of No.5 Squadron and was coded '7B-D' and stayed with the unit until April 1st 1951 when it joined No.612 Squadron, for a period of only fifteen days. After this brief sojourn came an even briefer one when RW388 was part of the Fighter Command Control and Reporting School from July 4th to the 9th 1951. On January 1st 1952 RW388 was the victim of a second flying accident and its flying days were finished abruptly. RW388 became instructional airframe number 6946M and was issued to No.49 MU at Colerne. It was moved to Colerne, still wearing the codes '3L-R'. After this it started a long career that was to keep the aircraft firmly in the public eye. It served in traditional Spitfire fashion on the gates at Benson, then Andover, before moving to No.5 MU at Kemble for a period of storage prior to moving to No.19 MU at St. Athan for 'conversion' to represent a Mark Vb Spitfire for the Royal Tournament. It was in No.19 MU's charge from April 10th 1968 and became for the exercise 'AB917'. On May 10th it had arrived at Cardington for the practice for the Tournament and had moved to Earls Court for the real thing in July. It moved from display at Earls Court to take part in the Edinburgh Tattoo in August and was displayed back at St. Athan for September and the station's Battle of Britain celebrations, after which it was placed in store with No.19 MU. (It should be said at this stage that the Spitfire had particular relevance in displays in 1968 as it was the year of the RAF's 50th Anniversary.) On October 6th 1969 the City of Stoke had stated its interest to display a Spitfire and RW388 was allocated to them, and preparations were put in hand to make ready the machine. The hold ups in getting the memorial ready meant a period of storage, firstly at No.71 MU, Bicester, from October 23rd, then at No.5 MU from December 12th 1970, finally returning to Bicester on March 8th 1971 for final touches to its now as original No.667 Squadron colours. RW388 was installed at Stoke on February 15th 1972.

43 *Hornet Moth G-ADOT at Stapleford 1955* George Nelson

42 *Mosquito TA634 in No.50 Hangar at Speke* P. H. Butler

44 *Fox Moth G-ACCB at Stratford* *MAPS*

45 *Spartan G-ACYK on the Hill of Stake* D. J. Smith

46 *Meteor G-ARCX on finals to Abbotsinch*

47 *W-2 in flight over Hanworth* 'Flight' 10698S

48 *Air-to-air near Farnborough of the SB.5 WG768, September 1953* *'Flight' 29018 S*

49 *'MACAWS' aerobatic team formation, with NEVVAA's Meteor WL181 leading* *Min. of Defence*

MOSQUITO AIRCRAFT MUSEUM

When in 1959 Mr. W. J. Goldsmith brought the historic prototype Mosquito back to its birthplace, little did he realise that the Museum he had created was the first of many such museums and societies that exist today.

It was in fact nineteen years earlier in 1940 that the 17th Century moated manor house of Salisbury Hall, once the site of a Roman Fort, and whose residents included such figures as King Charles II, Nell Gwynne, Sir Nigel Gresley and Sir Winston Churchill, first cemented its relations with the Mosquito when the De Havilland Aircraft Company took the house over, in order that its design team could work in secrecy on a high speed, unarmed, wooden bomber - the Mosquito.

Not only was the design work carried out at Salisbury Hall, but the first four prototypes were constructed there, in the grounds, the first being transported by road to nearby Hatfield, the remaining three using the fields adjacent to Salisbury Hall to take off from. The Mosquito went on to become one of the classic aircraft of World War Two, and excelled in all the roles it undertook, whether as a bomber, day/night fighter, photo reconnaissance, Pathfinder, high speed transport and many other uses which would take a book to list and describe!

Since the prototype Mosquito, W4050, arrived at Salisbury Hall it has been joined by a second Mosquito, a B. Mk.35 TA634, donated to the Museum in 1971 by Liverpool Corporation, the forward fuselage of B. Mk.35 TJ110 and, awaiting collection in Holland, the fuselage of FB. Mk.VI TA122. A selection of Mosquito armament and relics is also displayed.

Although based first and foremost on the Mosquito, the Museum is also dedicated to preserving as many other types of De Havilland aircraft as possible, thus being unique in being the only 'manufacturers' museum in Britain. Types that have been acquired include two marks of Vampire, Venom, Sea Vixen, HS.125, Tiger Moth, Hornet Moth, Chipmunk, Cierva C.24 and a Horsa glider, the last named together with the Vampire having also been designed at Salisbury Hall. Also on display is the White Collection of over 800 1:72 scale models.

The Mosquito Aircraft Museum is a voluntary, self supporting organisation and is registered as a charity, the proceeds of the entrance fees, donations etc. being given to the Royal Air Force Benevolent Fund. Although new additions are constantly being sought, the Museum's immediate aims are to build a second, much larger, hangar, and at the time of writing a Public Appeal for funds is about to be launched, with a target of £50,000.

A Supporters Society was formed in 1974 to help to maintain the growing collection and is undertaking several interesting projects, including the restoration of the Chipmunk, Tiger Moth, Hornet Moth and the fuselage of the Horsa glider. Any enthusiast is welcome to join, membership being £1.50 per year.

Salisbury Hall is situated near the village of London Colney, sixteen miles north of London and five miles south of St. Albans on route A6. Green Line coaches 713 and 714, buses 84 and 313 pass the main entrance. Admission for adults is 20p and children 5p. Free carparking, refreshments and sales stall are also available.

The Museum is open from Easter until September on Sundays from 1100 to 1800, and also on Thursdays July to September inclusive from 1400 to 1730 hours, Bank Holidays 1030 to 1730 hours. Parties at other times by arrangement.

Address : Mosquito Aircraft Museum, Salisbury Hall, London Colney, St. Albans, Herts.
Telephone : Bowmansgreen 23274 (contact Mr. Goldsmith or Stuart Howe).

Bleriot Type II No.54

Rebuild 1975 of original components, fitted with an Anzani 'V' twin of 25 horse power.

This aircraft was assembled from original parts as one of the last acts of the staff of the Museum at RAF Colerne. It was delivered to Salisbury Hall by road from there on April 13th 1976. This machine, and another at Cosford q.v., was donated to RAF Colerne by the late Leslie Goldsmith along with his unique civil registered, airworthy Bleriot Type XI G-AVXV on condition that the components were built into another aircraft. The identity quoted is thought to be the constructor's number for some of the components, though how relevant it is to the new airframes is unknown.

De Havilland DH.98 Mosquito W4050

Built 1940, development prototype, powered by two Rolls-Royce Merlin piston engines.

This aircraft was designed and built at Salisbury Hall, and was the prototype of the Mosquito 'Wooden Wonder' bomber. It was developed to be so fast that it needed no defensive arms. As E-0234 (De Havilland test registration) it first flew from the D.H. plant at Hatfield on

November 25th 1940 and was used throughout the War by the builders and other official users for test and development work, and for communications. In 1948 it was displayed at the Society of British Aircraft Constructor's display at Radlett and then moved into a period of store, firstly at Hawarden, and then at Hatfield, from 1953. It was then 'rescued' by Mr. Goldsmith and moved to Salisbury Hall on May 15th 1959 and rehoused in a small hangar bought with the proceeds of an appeal launched at the rime, built close to the site of the original hangar and design centre.For most of its War-time career it was camouflaged, but it has been repainted to the all over yellow colours in which it first flew.

De Havilland DH.98 Mosquito TT. Mk.35 TA634 'EG-F' see photo 42 and rear cover

Built 1945, target tug, powered by one Rolls-Royce Merlin 113 and one -114, piston engines, both of 1,710 horse power each.

TA634 was built at Hatfield in early 1945 as a B. Mk.35 bomber. It was issued to No.27 Maintenance Unit at Shawbury on April 14th and was to remain in store for the rest of the War and well beyond. In 1952 it was flown to Sywell for conversion to a target tug (TT. Mk.35) by Brooklands Aviation. After conversion it was sent to No.22 MU at Silloth for storage pending allocation to a unit. It first joined No.4 Civilian Anti-Aircraft Co-operation Unit at Llandow, only to end up in storage with the resident No.38 MU. It then moved to West Germany for a period with the Armament Practice Station at Sylt, followed by Ahlhorn APS and Schleswigland APS, until returning to No.27 MU in June 1957 for storage. It September 1959 it was delivered to No.3 CAACU and was coded '53'. It was to serve with No.3 CAACU at Exeter until being retired from active service in 1963. The work of the CAACU involved providing local TA units and the like with anti-aircraft practice, be it using blanks 'against' the aircraft itself, or using live ammunition against a target sleeve towed by the Mosquito. TA634 was retired to No.23 MU at Aldergrove in May 1963. It was bought by Liverpool Corporation for £720 and flown to Speke on November 6th of that year. The Corporation had plans to display the aircraft at Speke, but these were not to come to fruition, and TA634 again languished in a hangar. In 1968 it was picked for stardom in the film 'Mosquito Squadron' and was registered as G-AWJV for the purpose, this registration being allocated on May 31st. In the film TA634 masqueraded as a 'solid nose' fighter version and carried the serial 'HJ896' and the codings 'HT-G'. After the film TA634 returned to Speke (still carrying the fictitious film markings) and again became a hangar queen. With maintenance costs reaching intolerable levels the Mosquito's airworthiness was allowed to slip and in 1970 it was decided to donate the aircraft to the Mosquito Museum. It left Speke for London Colney by road on September 29th 1970. It has since been painted up in the colours of the aircraft flown by Group Captain P. C. Pickard, who led the 1944 Amiens prison raid, with the code marks 'EG-F'.

De Havilland DH.100 Vampire FB. Mk.6 J-1008

Built 1950s, single seat fighter, powered by one De Havilland Goblin turbojet of 3,350 lbs.

This aircraft was retired from active Swiss Air Force use in 1974 and donated to the Mosquito Museum. It was one of a large number of Vampire single seat fighters still in use with the Air Force mostly in the ground attack and operational training role as a back-up to the Venom fighter bombers that the force employs. These aircraft are only now scheduled to be withdrawn from use with the arrival of F-5E Tiger IIs in the early 1980s. J-1008 was flown direct from its last base, Dubendorf, to Hatfield on August 20th 1974 and maintained at Hatfield for a while, prior to space becoming available for it at Salisbury Hall. The Vampire was installed at the Hall on May 11th 1975 and is maintained in near airworthy condition, and there are plans to fly the aircraft again.

De Havilland DH.110 Sea Vixen FAW. Mk.2 XJ565

Built 1965, two seat all weather naval fighter, powered by two Rolls-Royce Avon 208 turbojets of 10,000 pounds static thrust.

XJ565 first flew on February 4th 1960, and served the Fleet Air Arm (as an FAW. Mk.1) with No.766B Squadron as'718/VL', and then No.893 Squadron. It was converted into a Mk.2 in 1965 and joined No.899 Squadron in 1967. It was retired on December 2nd 1968 with 1446 hours on the airframe. It was then used for arrester trials at the Royal Aircraft Establishment at Thurleigh. It moved by road to Salibsury Hall on October 31st 1976.

De Havilland DH.112 Venom NF. Mk.3 WX853

Built 1950, two seat night fighter, powered by one De Havilland Ghost 105 of 4,850 lb. s. t.

Built at Hawarden, this aircraft was awaiting collection on July 27th 1955, moving the following day across the airfield to No.48 Maintenance Unit. It was issued on October 5th to No.23 Squadron, the only unit with whom it was to serve. It was retired to No.27 MU at Shawbury on April 13th 1957 and declared a non-effective airframe on May 7th. It moved on to Debden for instructional use on July 10th and was allocated the maintenance airframe number 7443M. It served on the gate there until replaced by a Hunter and was moved to Salisbury Hall in January 1968 - still in its No.23 Squadron colours.

De Havilland DH.115 Vampire T. Mk.11 XD452

Built 1954, constructor's number 15266, advanced trainer, powered by one De Havilland Goblin 35 of 3,500 pounds static thrust.

Like WX853 above, XD452 was built at Hawarden, and was ready to be collected on June 30th 1954, but took some six days to be officially taken on charge by No.48 Maintenance Unit. It was issued to No.5 Flying Training School on August 16th, and did not receive a code. It was moved to Marshalls at Teversham on November 16th 1956 and stayed there until April 30th the following year. At the end of this work it was allocated to No.10 MU at Hullavington on May 13th, moving on to No.22 MU at Silloth on January 5th 1958. It must have then received damage as it is recorded as having received the attentions of a crew from No.60 MU Church Fenton from March 2nd until March 31st 1959. It was on charge with No.12 MU at Kirkbride from May 26th 1959, and with No.22 MU again from June 14th. It re-joined No.5 FTS on June 30th and became 'AJ', and later '31'. It was returned to the makers for re-finishing on March 12th 1960 and was allocated next to No.8 FTS from April 13th 1962, coded '31' still. It was back with the makers for modifications from September 18th until December 6th 1962, returning to No.8 FTS that day. XD452 was then involved in a Category 3 flying accident on April 1st 1963 and was again in the hands of a crew from No.60 MU from April 3rd to May 22nd. It was back in service with No.8 FTS from the following day and then went on to 'do the rounds' of other FTS's, all with the code '66', as follows :- No.1 from February 17th 1964, No.7 from January 3rd 1966 and No.3 from November 4th 1966. It was retired to No.27 MU at Shawbury on September 1st 1967 and declared non-effective stock on October 16th 1967. It was allocated the maintenance serial 7990M and was moved to London Colney on February 19th 1968.

De Havilland DH.115 Vampire T. Mk.11 XE985 '64'

Built 1954, constructor's number 15572, other data as XD452.

This aircraft as displayed is a hybrid, being the booms and cockpit of XE985 and the wings of WZ476. This mixture was acquired from Hawker Siddeley Aviation at Woodford and delivered by road to Salisbury Hall in mid 1975, but not assembled and displayed until the 1976 season. Both components had been held in store at Woodford in the hope of possible resale, and both aircraft had been at one time bound for the Chilean Air Force, this deal however fell through. The aircraft were then offered by HSA to preservation societies and the like, the XE985/WZ476 combination being one of the last of many such Vampires to leave Woodford for all parts of the country. XE985 was built at Hawarden and was last in use with No.5 Flying Training School at Oakington, where it was coded '64'. It was bought back off the RAF on November 17th 1967 and stored initially at Hawarden, moving to Woodford in the winter of 1971/2. WZ476 was also built at Hawarden, constructor's number 15089, and last served with the Royal Air Force College at Cranwell, at one time being coded '52'. It was bought back, at St. Athan, on February 21st 1968.

Notes : The following aircraft are under active restoration in the Museum's workshops and normally viewable on the Sunday openings, or by appointment.

Airspeed AS.58 Horsa II 1943 troop/vehicle carrying assault glider, fuselage section, ex-Banbury, donated by Marilyn Pettifer. Plate bearing 'HHM317' inside.
De Havilland DH.82A Tiger Moth G-ANFC c/no. 85385, built by Morris Motors as DE363. Arrived by road November 2nd 1975 from Dunstable. For restoration to military status.
De Havilland DH.82A Tiger Moth G-ANFP c/no. 82530, built by De Havillands as N9503. Used for spares at Rush Green 1963 and discovered as a bare frame at Denham. Owned by S. Howe.
De Havilland DH.98 Mosquito TT. Mk.35 TJ118 Forward fuselage, discovered at Elstree Film Studios. Rear end under totally separate restoration in Oxford.
De Havilland DHC-1 Chipmunk T. Mk.10 WP790 c/no. C1/0682, last served with No.8 AEF at Shawbury, coded 'T'. Civilianised as G-BBNC, but used only for spares, donated to the MAM.

The following aircraft are in store, or in restoration, or are not at the time of writing available for public inspection. Most are located at sites away from the Museum :-

Cierva C.24 G-ABLM De Havilland built, c/no. 710, built at Stag Lane in 1932, large cabin autogiro. In store, completely refurbished at Hatfield, on loan from the Science Museum.

De Havilland DH.82A Tiger Moth G-ANRX c/no. 3863, built by De Havilland as N6550. Withdrawn from use as Boxted in 1961. Arrived in mid 1976 from Belchamp Walter. In store at Salisbury Hall.

De Havilland DH.87B Hornet Moth G-ADOT c/no. 8027, built 1935, impressed 29/ 2/40 as X9326. Returned to G-ADOT in 1946 until withdrawn in 1959, under restoration at Hatfield, acquired via the Shuttleworth Trust, was a child's plaything in Southampton. *see photo 43*

Hawker Siddeley HS.125 Series 1 G-ARYC c/no. 25003, first production example, used for all its life on engine work with Bristol Siddeley/Rolls-Royce. Rescued from dereliction at Filton and moved to Hatfield in March 1976. In store there.

MOTIVE POWER MUSEUM

The Motive Power Museum is a modern, purpose-built, building designed to portray as large a spread as possible of the development of power-driven transport. There is an emphasis on steam-driven transport, especially locomotives, but aviation is not forgotten, with a Tutor glider and a Vampire and a Provost on display. The aircraft were initially displayed externally (apart from the Tutor which has always been housed in the main hall), but have since been brought under the wing of a new extension.

The Museum is in Dock Road, Lytham St. Annes, just off the A584 Preston to Blackpool road. Opening times for 1977 are not available to the publishers, but are thought to be 1100 to 1700 hours Tuesday to Sunday, the Museum not being open on Mondays. The Museum closes for the winter, opening again at Whit/Easter. 1977 admission prices are not available. Further details are available from the Museum.

Address : The Motive Power Museum, Dock Road, Lytham St. Annes, Lancashire.

De Havilland DH.115 Vampire T. Mk.11 XK624 '32'

Built 1956, constructor's number 15781, advanced trainer, one De Havilland Goblin 35.

Built at Hawarden, XK624 was awaiting collection on July 13th 1956 and was delivered to No.7 Flying Training School three days later. It was involved in a Category 3 (Repair beyond unit capacity) flying accident on June 13th 1957 and was repaired on site by a team from No.54 Maintenance Unit, based at Cambridge, from June 26th until August 16th, re-joining No.7 FTS that day. This was short-lived, as it went Category 3 again on November 4th. This time De Havillands were called for to repair it, this taking place from December 9th until February 5th 1958. It was back in service with No.7 FTS the following day. It was refurbished by the makers from September 22nd 1958 until June 19th 1959, going back to No.7 FTS on July 2nd 1959. It was briefly loaned to the Ministry of Aviation from September 1st 1960, but was on charge with the Central Flying School at Little Rissington from August 1st 1961 for a spell of only nine days. On August 10th it was taken on charge by No.23 Group Communications Flight and given the code '44', later changed to '79'. It joined No.1 FTS, still as '79' from October 13th 1965, going back to serve with No.7 FTS from January 14th 1966, this time coded as '52'. On November 4th that year it was transferred to No.3 FTS and finally back to the CFS on January 16th 1967, coded '32'. It was acquired by the Museum and flown into Squires Gate in December 1971, for removal by road to Lytham.

Percival P.56 Provost T. Mk.1 WW421 'P-2'

Built 1954, constructor's number P.56/234, basic trainer, powered by one Alvis Leonides 126 radial piston engine of 550 horse power.

WW421 was awaiting collection at Luton on September 24th 1954 and was issued to No.12 Maintenance Unit at Kirkbride five days later. It joined No.3 Flying Training School as 'P-B', later 'Q-H', on October 21st and served there until being retired to No.27 MU at Shawbury on May 1st 1958. It was declared as non-effective stock on June 28th 1960 and allocated to No.4 School of Technical Training on October 27th, maintenance serial 7688M. It later moved from No.4 SoTT at St. Athan to No.1 SoTT at Halton, here it is thought to have been cross-bred with WW450/7689M and given the latter's wing. This may have occured at St. Athan. It moved to Lytham in 1971.

Slingsby T.8 Tutor BGA.833

Built early 1950s, trainer glider.

This aircraft was built as a Cadet TX. Mk.2 for the Royal Air Force as VW535. It was civilianised as British Gliding Association certificate number 833 in September 1957 as a Tutor. It is not known where the Museum acquired this aircraft from, possibly locally. It is displayed in the Main Hall.

MUSEUM OF ARMY FLYING

The Museum is located at the Army Air Corps Centre at Middle Wallop, about six miles from Andover on the A343 Salisbury Road. It is open to the public and there is no charge for admission. Normal opening hours are Monday and Wednesday from 1000 to 1200 and 1300 to 1600; Friday 1300 to 1600 hours. In addition the Museum is open on special days and special viewing times may be arranged on application to the curator. Visitors are invited to park their cars on the Army Air Corps Officers Mess car park on the other side of the A343, opposite the Museum.

The Museum was opened in its present location in 1974 and provides a record of all aspects of army flying. There are many interesting exhibits including aircraft which have been in service with the Army. There is a library of books and films together with records of the history of the air observation post squadrons, the Army Air Corps and some aspects of the history of the Glider Pilot Regiment. These records include accounts of the operations of squadrons in the Second World War, Korea, Malaya, Suez, Cyprus, Aden, Borneo, and Northern Ireland. There are many original pamphlets, documents, scrapbooks and photographs, some of which are very delicate and irreplaceable, thus the library is not open to the public for general viewing, however, particular items, may be seen by prior permission with the curator. Although space in the present building is severely limited, the fuselage section of a Horsa glider, an Auster Mark 9 aeroplane, and a Skeeter helicopter are on display.

A number of other aircraft are on display at the Army Air Corps Centre which is an active flying station where all the Army pilots receive their training. During a visit to the Museum the Sioux, Scout and Gazelle helicopters are likely to be seen on training flights.

The Museum of Army Flying is still in the formative stage and is supported entirely from voluntary subscriptions and donations. A small income is derived from the sale of certain badged souvenir items in the foyer of the Museum. New exhibits for the Museum are always welcome and it is possible that you have items or documents; perhaps a photograph or just information which would make a valuable contribution to an exhibit, display or the archives. If so you are invited to contact the curator, who is always available to answer queries concerning particular aspects of army flying or Army Air Corps history.

Address : Lt. Col. L. L. S. Williams R.A. (Retd.), Curator, Museum of Army Flying, Army Air Corps Centre, Middle Wallop, Stockbridge, Hampshire SO20 8DY.

Telephone : 0264 62121.

Airspeed AS.58 Horsa II BAPC.80

Composite reconstruction, parts built 1944, troop or vehicle carrying assault glider.

This forward fuselage has been painstakingly rebuilt from several different components by the Museum. This rebuild has been allocated the British Aircraft Preservation Council number 80 for identification purposes. It is thought that the largest component used was that of Harris Lebus built TL726, which was originally the property of the now-defunct Historic Aircraft Preservation Society.

Auster B5 AOP. Mk.9 WZ721

Built 1951, aerial observation post, powered by one Blackburn Bombadier 202 in-line piston engine of 173 horse power.

This aircraft was originally with No.6 Flight, Army Air Corps, and then entered a period of storage with No.19 Maintenance Unit at St. Athan. On September 15th 1960 it left the Army Air Corps Centre at Middle Wallop on a flight to Singapore on delivery to No.656 Squadron. This notable flight for so diminutive a craft took it via Lyon, Salon, Malta, Idris, Cairo, Khartoum, Asmara, Jiwani, Rangoon and Taiping - saying nothing for the range of the Mark 9, but gallons for the crew! In 1965 or 1966 it became part of No.4 Royal Tank Regiment at Paros, Serembam, in Malaya, where it was christened 'Dragon'. By November 1966 it was with No.389 MU at Seletar, Singapore and up for sale. By July 1973 its fuselage was noted at Middle Wallop and it joined the Museum in 1974.

Saunders-Roe Skeeter AOP. Mk.12 XL813

Built 1959, aerial observation post, powered by one De Havilland Gipsy Major 215 of 215 h.p.

XL814 was ready for collection by the Army Air Corps at Eastleigh on April 29th 1959, but was briefly loaned to the builders in the following September to enable it to appear at the Society of British Aircraft Constructor's display at Farnborough. It was issued to No.15

Maintenance Unit at Wroughton on May 31st 1962. On July 20th it joined the Advanced Rotary Wing Flight at Middle Wallop, but returned to No.15 MU on October 23rd 1964. On June 11th the following year it was flown to Detmold in West Germany and in September joined No.9 Flight based at Soltau. By November 1967 it was with No.4 Field Regiment Royal Artillery, yet at the same time was reported to be at RAF Halton! Now its official history gets still more confusing with the note that it joined the Museum of Army Flying on January 23rd 1968, yet was not at Middle Wallop until March 25th. Clarification welcomed.

Note : Displayed at the main gate of the Army Air Corps Centre, very close to the Museum of Army Flying are the following :-

Auster B5 AOP. Mk.9 WZ670 This aircraft was put on the gate in 1968 and is really WZ724. It has been painted to represent an aircraft used during the Malayan Emergency.
Bristol B.171 Sycamore HR. Mk.14 XG502 Joined the gate in 1972, painted to commemorate the aircraft of No.1906 Light Liasion Flight RAF and of the Joint Experimental Helicopter Unit.
Saunders-Roe Skeeter AOP. Mk.12 XL769 Put on the gate in 1971, is a hybrid, and not what it says it is! It is made from the fuselage of XL738 and the tail boom of XM565.

MUSEUM OF ARTILLERY

The Museum of Artillery can trace itself back to 1778 when Captain William Congreve R.A., a famous artilleryman, opened to exhibition models that he had made from 1776 in the 'infant museum' as he called it. His son, Colonel William Congreve, was primarily responsible for causing the Rotunda to be erected in 1820 on its present site. The first Congreve went on to form the 'Military Society' and, thanks to the support of King George III, had a Repository built at Woolwich from 1778. This was at first called the Warren, and later The Arsenal. The Warren was gutted by fire in 1802 and the museum pieces moved to temporary premises. In 1819 one of the six double-ceiling tents designed by John Nash for the Meeting of Allied Sovereigns was erected on the present site and used for the museum. It was opened on May 4th 1820 and was later made more permanent by the original designer. In 1870 the administration of the Rotunda was transferred to the Royal Artillery Institution, who are still responsible for its running.

The collection has expanded considerably and exhibits include a 15" wrought iron Bombard from Bodiam Castle in Sussex - this is one of the oldest pieces of ordnance in existence, a 13 pounder QF gun from the Battle of Nery September 1914, two Russian bronze guns used in the Crimea, a 1945 self-propelled gun, and, a recent addition, an Auster air observation post gun spotter from the Army Air Corps.

The Museum opens in two periods : Winter Period : Weekdays and Saturdays 1000 to 1245 and 1400 to 1600, Sundays 1400 to 1600 hours ; Summer Period : Weekdays and Saturdays 1000 to 1200 and 1400 to 1700, Sundays 1400 to 1700 hours. The Museum is closed on Christmas Day and Good Friday. Admission is free.

Address : The Museum of Artillery, Royal Artillery Institution, Old Royal Military Academy, Woolwich, London, SE18 4JJ.

Telephone : 01-854 2424.

Auster B5 AOP. Mk.9 XR271

Built 1961, constructor's number AUS/B5/188 (unconfirmed), aerial observation post, powered by one Blackburn Bombadier 202 in-line piston engine of 173 horse power.

XR271 flew for contractor test flights on March 29th and 30th 1962, having been delivered to the Army Air Corps on the 6th. It flew its acceptance check on April 13th, at Middle Wallop, being delivered there on April 6th. It flew with the Advanced Fixed Wing Flight. It was delivered to No.19 MU at St. Athan on January 16th 1967, it had totalled 1,070 flying hours. It was transferred to Larkhill on March 2nd 1970 and thence to the Museum.

MUSEUM OF FLIGHT

The Museum is an extension of the Department of Technology of the Royal Scottish Museum in Edinburgh (which see). It is situated approximately 20 miles due east of Edinburgh and occupies part of the site of the former RAF wartime station at which night fighter aircrew were trained. East Fortune also has a place in history as the departure point in July 1919 of the airship R.34 on the first east/west and first double crossing of the Atlantic by air.

The birth of the Museum can be said to have taken place in 1971 when the Royal Scottish Museum acquired a Spitfire XVI from the Ministry of Defence and obtained permission to house it in a corner of a hangar at East Fortune then used as a Government store. Before long the complete hangar became available giving room for expansion and an adequate area for presentation. Public visits commenced in 1973 and the collection has grown steadily, the aircraft on display being augmented by aero engines ranging from a 1909 Green to a Concorde Olympus. A section of the Museum is also devoted to rockets, pride of place being taken by Blue Streak. Displays are planned to give visitors not only an outline of the history of aircraft and rockets but also an insight into modern technology. A further hangar has recently been acquired which will permit continued development of the Museum.

Admission free. Open to the public from 1000 to 1600 hours on 3rd April, 1st May, 5th June and 2nd October 1977 and also daily during July and August. Open at other times to societies and groups by prior arrangement.

Address : Museum of Flight, East Fortune Airfield, North Berwick, East Lothian EH39 5LF.
Telephone : Athelstaneford 308.

Auster AOP. Mk.5 'TJ472'

Built 1940s, aerial observation post, powered by one Lycoming O-290 piston engine.

This exhibit is on loan from the Aircraft Preservation Society of Scotland, and moved to East Fortune in 1976, from its previous home at Inverkeithing, near Edinburgh. This aircraft was in use for many years as a frame for technical training with Airwork at Scone, and before that with Air Service Training at Hamble. No marks have been found to identify it exactly. It is known to be an AOP. Mk.5 frame. At one time at Scone, it wore the bogus registration G-ALES. It was bought from Airwork by APSS and removed to Inverkeithing where it was refurbished and finished in the colours of a typical AOP. Mk.5 of the War years, and given the typical serial TJ472. The British Aircraft Preservation Council have allocated this airframe the identification number 70.

British Aircraft Swallow II G-AEVZ

Built 1937, constructor's number 475, two seat light aircraft, one Blackburn Cirrus Minor.

G-AEVZ was delivered to the Hull Aero Club early in 1937 and served with the club, at Hedon, until the War, when it was stored at Gainsborough. It remained dormant until July 1954 when the aircraft was registered to E. Clark of Doncaster. In October 1958 it moved home, not very far away, to Yeadon with a Mr. D. Pullen. In March 1964 it was in a dismantled state at Crosby-on-Eden airfield, near Carlisle. In February of the following year it was donated to the Northern Aircraft Preservation Society, who moved it down to Manchester and undertook a refurbishing programme on it. It appeared at a few local displays before being loaned (by the now renamed Aeroplane Collection) to the Museum of Flight.

BAC 221 WG774

see photos 39 & 40

Originally built 1952, ogee-wing aerodynamic research aircraft, one Rolls-Royce Avon.

The BAC 221 is a complex test airframe designed to evaluate the ogee wing of the Concorde supersonic airliner. It was based on the airframe of the record-breaking Fairey Delta 2 and converted in a dramatic way to the new test role. This employed a new wing structure and many other changes, including completely new systems and telemetry and a new long-stroke undercarriage. The conversion was undertaken at BAC's Filton plant. The FD-2 (as the aircraft originally was) was built also for a research mission, this time into high speed flight. It was built at Hayes to Specification ER.103, and was given the Fairey Type V. Construction was on a low priority and the sleek aircraft was not finished until 1954. It flew for the first time on October 6th 1954 from Boscombe Down. On March 10th 1956 it took the World Speed Record with a speed of 1,132 m.p.h. It was retired after only a short flying career and earmarked for conversion to BAC 221 status, flying in this guise in 1964. It was employed by the Aerodynamics Flight at the Royal Aircraft Establishment until moving to join the collection at East Fortune in 1975.

Beech E.18S G-ASUG

Built 1956, constructor's number BA-111, general purpose transport, powered by two Pratt and Whitney Wasp Junior R985-AN-1-14B radial piston engines of 450 horse power each.

This aircraft has considerable connections with Scotland - being used for most of its life in the United Kingdom with the local airline, Loganair. G-ASUG was acquired by the airline in May 1968 and was to be used basically for charter work, although it later took on some scheduled work as well. Indeed G-ASUG undertook Loganair's first international service, Abbotsinch-Dyce-Stavanger from July 1969. In its later days G-ASUG was kitted out with an executive interior. It was withdrawn from use by Loganair in March 1975 at their Abbotsinch base. In 1976 it was moved to East Fortune and presented to the Museum. Prior to its service in Scotland G-ASUG had put in a fair bit of flying in the United States. It was built by the Beech Aircraft Corporation as N575C and later became N555CB and then N24R.

De Havilland DH.89 Dragon Rapide G-ADAH

Built 1935, constructor's number 6279, light feederliner, powered by two De Havilland Gipsy Six in-line piston engines of 200 horse power each.

G-ADAH serves as a good contrast to G-ASUG above, as it was to serve in exactly the same role in Scotland before the Second World War as the Beech was to serve after it. G-ADAH was built at Stag Lane in 1935 and first saw service with Hillman's Airways Limited.
In February 1936 it changed ownership, but not base, to British Airways Limited until being sold in August. That sale saw it move for the first time to Scotland with Northern and Scottish Airways, based at Renfrew. Exactly a year later it was registered to Airwork Limited down at Shoreham, but was back in Scotland by October of the following year when it joined the pioneering Allied Airways (Gandar Dower) Limited, based at Dyce. It was with Allied that G-ADAH gained the name 'Pioneer'. G-ADAH still wears the livery from this period to this day. After a long period of inactivity and store 'Pioneer' appeared at Booker in 1969 and was acquired by the Northern Aircraft Preservation Society (now The Aeroplane Collection) via the Shuttleworth Trust in October 1970. It was moved to Peel Green, Manchester and stored. In 1975 it was loaned to the Museum and moved north by road.

De Havilland DH.112 Sea Venom FAW. Mk.22 WW145

Built 1955, two seat all weather fighter, powered by one De Havilland Ghost turbojet of 4,850 pounds static thrust.

WW145 was built at Christchurch and first served with No.891 Squadron, coded '435/O'. It was built as an FAW. Mk.21 version, and was converted to FAW. Mk.22 status by December 1st 1958. It re-joined No.891 after conversion and remained with them until January 1965 when it became part of No.750 Squadron, coded '591/HF', based at Hal Far, Malta. It was at the Royal Naval Aircraft Yard at Sydenham from July 1965 and was further modernised until 1968 when it was re-issued to No.750 Squadron, then at Lossiemouth for spares and fire practice use. It is recorded as having been reduced to spares and produce by February 1970. It is assumed to have been held at Lossiemouth until acquired by the Museum in 1972/3.

De Havilland DH.115 Sea Vampire T. Mk.22 XA109

Built 1954, advanced trainer, powered by one De Havilland Goblin 35 of 3,500 lbs. s. t.

XA109 was first on charge with No.802 Squadron from about February 1954, moving on to No.781 Squadron by July 1955. In 1956 it was with the Junior Officer's Air Course and was painted in mock BOAC colours to emphasise the similarity of the unit's initials to that of the country's international state airline. It next briefly served with No.831 Squadron as a 'hack' and continuation trainer from June to August 1962, as '990/CU'. It was back with the unit by September 1963, this time as '385'. It is next recorded as going to the Aircraft Holding Unit at Lossiemouth and being reduced to spares and produce - the remains joining the fire heap. Like WW145, above, it is thought to have lingered on untouched despite the official death sentence and to have been saved by the Museum.

Gloster Meteor 14 G-ARCX

see photo 46

Built early 1950s, two seat avionics test bed, powered by two Rolls-Royce Derwent 9 turbojets of 3,600 pounds static thrust.

G-ARCX was built originally by the Armstrong Whitworth factory at Baginton as a NF.11 night fighter. It was chosen by Armstrong Whitworth's at Baginton to be the prototype for their NF. Mk.14 night fighter variant. It first flew as such on October 23rd 1953 from Baginton, still as its original identity, WM261. It spent most of its service life on manufacturers trials. In September 1960 it was allocated to Ferrantis, who had a flying unit at Turnhouse. After a long period of modification to enable it to become a test-bed for a large number of avionic trials the machine flew again from Turnhouse on January 21st 1963, as G-ARCX. Its civil registration being required for the nature of some of the tests the aircraft was to carry out. After a long and successful test career G-ARCX was retired from active service in February 1969 and placed in store at Turnhouse. In 1975 it was donated to the Museum of Flight and moved to East Fortune.

Hawker Sea Hawk F. Mk.1 WF259 '171/A'

Built 1954, single seat naval fighter, powered by one Rolls-Royce Nene turbojet.

WF259 was taken on charge by the Fleet Air Arm at Stretton on March 11th 1954. It was issued to No.807 Squadron at Brawdy on May 7th, but moved on to No.802 Squadron by July. It was coded '124' and carried the unit's black cougar motif on the nose during this time.

It returned to Stretton on March 12th 1955, this time with a unit, No.767 Squadron. It was at Fleetlands from May 1955 until January 1956 at least, and at Abbotsinch by July 1956. On November 22nd 1956 it joined No.700 Squadron at Ford, as '511/FD'. It was delivered to Lossiemouth on February 8th 1957 and was taken on charge with No.738 Squadron on the 27th, becoming '638/LM. It was transferred to No.736 Squadron, still at Lossiemouth, on March 10th 1957, and eventually moved on to the Station Flight before being declared an instructional airframe in July 1959. It was allocated the maintenance serial A2483. It was acquired by the Museum and has been painted in ficticious codings.

Messerschmitt Me 163B-1 Komet 191659 '15'

Built 1944, single seat interceptor, powered by one Walter 109.509 bi-fuel rocket motor.

This aircraft was captured in Germany in 1945 and was transferred to the Royal Aircraft Establishment at Farnborough with the Air Ministry Number 215. Soon after the War it was handed over to the College of Aeronautics at Cranfield for use in their 'Library of Flight' and was part of their exhibition for many years. It was acquired by the Museum in 1975 and was delivered by road to Teversham for refurbishing by Marshalls on November 4th 1975, moving to East Fortune in early 1976.

Mignet HM.14 Pou du Ciel BAPC.12

Built 1930s, single seat ultra-light, powered by one Scott A.2S Flying Squirrel of 25 h.p.

This 'Flea' was built by S. O. Whiteley in Halifax at the height of the Flea craze. He had a converted Austin 7 motor-car engine to power it. The aircraft, unregistered, was tested at Rishworth, near Oldham, it is not known how well (or otherwise) it flew. It was put into store until June 1964 when it was taken on by the Northern Aircraft Preservation Society (now The Aeroplane Collection). It was allocated the British Aircraft Preservation Council identity number 12. NAPS re-engined it with a more-conventional Scott, as originally used on HM.14 G-ADYO. From 1974 it was loaned to the Newcastle Museum of Science and Technology. In early 1976 it was loaned to the North East Veteran and Vintage Aircraft Association and displayed at the group's Museum at Lambton Park, Chester-le-Street. It was delivered by road to East Fortune on loan on September 26th of that year.

Slingsby T.12 Gull 1 BGA.902

Built 1938 or 1939, single seat glider.

The previous identity and the constructor's number of this glider are unknown. It was one of ten of the type built in 1938 and 1939. A Gull made history as the first sailplane to fly across the English Channel on 22nd April 1939 - this flight being made by Mr. G. Stephenson. The cross-Channel Gull is believed to have been BGA.379, and it is said that parts of this aircraft are included in the mysterious BGA.902. BGA.902 had its first certificate of airworthiness issued in May 1959, its first recorded owner being the Aberdeen Gliding Club; it was later sold to D. Heyhurst before being handed over to the Museum in 1971.

Saunders-Roe Skeeter AOP. Mk.12 XL762

Built 1958, constructor's number S2/5074, aerial observation post, powered by one De Havilland Gipsy Major 215 of 215 horse power.

XL762 is recorded as being ready for collection at Eastleigh on September 11th 1958. Its introduction into service was briefly interupted so that it could appear as a SARO exhibit at the Society of British Aircraft Constructor's Display at Farnborough that month. It was first issued to a unit on the 16th, No.651 Squadron. In January 1959 it suffered an accident and from March 1959 until January 1960 it spent much time shuttling back and forth from its unit to the makers whilst all was ironed out on it. On February 7th 1964 it was moved to No.654 Squadron. From late 1965 it served with No.9 Flight and then joined the 2nd Royal Tank Regiment, based at Hohne, in West Germany, during 1966/7. On November 23rd 1967 it was delivered to No.15 Maintenance Unit at Wroughton for disposal or re-allocation. By July 2nd of the following year it was allocated the maintenance airframe number 8017M, having been declared a non-effective airframe in the previous May. It joined the Air Engineering Training Wing at Middle Wallop shortly afterwards. In November 1975 it was moved to East Fortune by road.

Spartan Cruiser III G-ACYK see photo 45

Built 1935, constructor's number 101, feederliner, powered by three De Havilland Gipsy Major in-line piston engines of 130 horse power each.

Although only a gutted forward-fuselage, this exhibit is a unique relic of a rare pre-War short-range airliner. Whilst operating with Northern and Scottish Airways Limited (with whom it had been since August 1936) G-ACYK crashed into the high ground of the Hill of Stake, near Largs, in Scotland, on January 14th 1938. From that date until 1973 the wreckage remained. Much was salvaged with the finding of the wreck, but much more was to stay on site for many years. The fuselage became a well-known land-mark for fell-walkers, and it has taken its share of people chalking (or worse) their names (or worse!) on it. By 1973 only the forward fuselage remained and the Royal Scottish Museum and other interested bodies attempted its salvage. The fuselage had never been moved before because of its bulk and the difficulty of the terrain in which it was lodged. The new salvage bid used a Sea King HAS. Mk.1 helicopter to lift the fuselage away on July 25th 1973. The fuselage has since been displayed at East Fortune and is a further example of the Scottish airline scene for the Museum. G-ACYK was originally registered to Spartan Airlines on April 16th 1935, it served with British Airways Limited from the April of the following year, before joining Northern and Scottish.

Supermarine Spitfire LF Mk.XVI TE462

Built 1945, single seat fighter, powered by one Rolls-Royce Merlin piston engine.

TE462 was first issued to No.39 Maintenance Unit at Colerne on June 26th 1945 and was to remain stored for what was left of the War and beyond, being moved to No.33 MU at Lyneham on February 22nd 1950. It was issued to the Station Flight of Finningley on October 11th 1950 and suffered a Category 3 flying accident (repair beyond unit capacity) on April 23rd 1951. On September 6th 1951 it was on charge with No.101 Flying Refresher School and served with them until being delivered to No.29 MU at High Ercall to bring to a close a very brief flying career. It was declared a non-effective airframe on December 14th 1954 and was given the maintenance airframe number 7243M. It was issued to Ouston on August 19th 1955 and displayed on the gate there, with the codes 'RAN-B'. It then moved to Kemble (on August 18th 1970) on Ouston's closure and then to No.71 MU at Bicester on December 9th 1970 for finally preparation for the journey by road to East Fortune, which was undertaken on February 17th to 19th 1971.

Weir W-2 'W-2'

see photo 47

Built 1934, single seat autogyro, powered by one Weir in-line piston engine.

The firm of G. & J. Weir have been pioneering autogyros and helicopters for many years, under one aegis or another. Their initial experiments are recorded with this sole surviving relic of their early days, the W-2. It was built at Cathcart in 1934, and flew for the first time in May. In 1936 it was fitted with the tail surfaces and the rotor pylon of the planned W-3 development, and it still wears these today, giving it a hybrid look. It test-flew as W-2, Weir's prefix being 'W-'. After the War it was acquired by the Science Museum and stored, possibly in the later days at Hayes. It was loaned to the Museum of Flight as being of special significance to Scotland. It has been allocated the British Aircraft Preservation Council identity number 85.

Note : The Museum of Flight is at present holding the following aircraft, which may join the collection :-

SAAB S.91D Safir G-ANOK Built 1954, c/no. 311, allocated SE-CAH before delivery but not taken up. Until 1975 was resident at Turnhouse.

NEWARK AIR MUSEUM

The Newark Air Museum really started in 1963 when Charles Waterfall and Neville Franklin discovered the remains of the Westland Wallace at Cranwell. These were removed to Newark in 1965.

Other items including the Monospar, Magister, Flying Flea and Prentice were gathered together, and fortunately a permanent site was leased from the Newark Agricultural Society on their showground in 1970. This is on the ex-RAF airfield at Winthorpe three miles from Newark on the A46 in the direction of Lincoln.

Since then there has been a steady flow of additional material, and there is now a display of complete aircraft as well as a hangar containing a display of engines and many smaller items. The site was enlarged in 1975, and is currently being further enlarged.

The Museum is a founder member of the British Aircraft Preservation Council, and is a strictly non-profit making organisation dependent entirely upon membership subscriptions, donations, gate fees and shop sales. It operates as a limited comapny and is a registered charity.

Membership is now around the two hundred mark, and the Museum organises a programme of

50 *B-25 N9089Z outside the Surrey and Kent hangar at Biggin Hill, prior to joining the HAM*

P. H. Butler

51 *Newark Air Museum's Varsity touches down at Winthorpe from Finningley April 1st 1976*

P. H. T. Green

52 *Hendon July 1951 : Cygnet G-EBMB, Desoutter G-AAPZ, Fulmar G-AIBE* 'Flight' 25985 S

53 *Sunderland ML824 on its way to Pembroke Dock, March 23rd 1961* 'Flight' Se 20614 2

54 *The RAF Museum's Star Meteor EE549 at Tangmere, August 1946* 'Flight' 19491 S

film shows and lectures during the winter months.

Opening times are from 1400 to 1800 on Sundays, April to October, and on Bank Holidays during that period. Admittance for 1977 is 30p for adults and 10p for children.

For other information please apply to the Hon. Secretary at 141, Hawton Road, Newark, Nottinghamshire, NG24 4QG.

Avro 652A Anson C. Mk.19 VL348

Built 1948, general purpose transport, powered by two Armstrong Siddeley Cheetah 15s of 420 horse power each.

This aircraft had a fairly uneventful career whilst serving with the Royal Air Force (as indeed many 'Annies' had) in the communications role. It served on the Communication Flights of the following RAF Groups :- Nos. 22, 24, 62 (Home Command) and 81 - the latter based at Colerne. Its service life came to an end in 1966/7 and the aircraft was stored at No.27 Maintenance Unit at Shawbury. It was sold to T.A.T. Limited and registered to them on November 6th 1967 as G-AVVO. In temporary ferry markings she was flown to Halfpenny Green airfield four days later. T.A.T. did not convert G-AVVO for civil use and she was eventually bought by the British Historic Aircraft Museum at Rochford. G-AVVO was moved there by road on September 3rd 1968. With the misfortunes of starting that museum, the aircraft began to decay in open storage. By 1971 the Historic Aircraft Museum had been successfully formed and G-AVVO was in use as a spare airframe, as that museum had a rarer Anson on display (G-AGPG). Newark Museum suffered the terrible loss in 1971 of the burning by vandals of their Mark XI Anson G-ALIH. To replace this machine Newark acquired G-AVVO from the HAM in the November of 1971. The outer wings of G-AVVO has suffered most, and the wings of G-ALIH (which thankfully were unharmed) were mated in their place. The machine is now displayed in its original military hue as VL348 once again.

Bristol B.171 Mark 4 Sycamore HR. Mk.14 XE317 'S-N'

Built 1956, constructor's number 13198, general purpose helicopter, powered by one Alvis Leonides radial piston engine of 520 horse power.

XE317 was laid down on the production line at Filton for the civilian market, registered as G-AMWO. It was not used in this way however, and entered service with the Royal Air Force as XE317. It served with the Central Flying School at Ternhill, where it was coded 'S-N' - the coding that it still wears today. XE317 was retired in the late 1960s and was transferred to the Royal Navy, who used it for many years in the training of helicopter pilots on how to leave a ditched machine, and for training rescue frogmen. For this purpose it was dunked in a water tank at H.M.S. 'Vernon', Portsmouth. It was retired from this unusual role in 1975, having been replaced by a purpose-built mock-up, representing a more modern service helicopter. It was then bought by the Museum, and delivered to Winthorpe by road on August 8th 1975.

De Havilland DH.115 Vampire T. Mk.11 XD593 '50'

Built 1954, constructor's number 15538, advanced trainer, one De Havilland Goblin 35.

XD593 was built at Hawarden and was handed over to the Royal Air Force on September 30th 1954, being placed in the care of No.10 Maintenance Unit at Hullavington on October 5th. Its first active service came with No.4 Flying Training School from October 29th. It moved on to No.5 FTS at Oakington on March 9th 1955. On the 15th February 1957 it again moved, this time to the Fighter Weapons School at Leconfield where it was involved in gunnery training. It returned to No.10 MU on May 1st 1957, and moved on to No.5 MU at Kemble on March 3rd 1959. In 1960 it was put back into service, this time with the Central Flying School at Little Rissington on 10th October 1960. XD593's last service came from December 19th 1962 with No.8 FTS at Swinderby. Here it acquired the code '50'. Nearly a year later (December 13th 1963) it was flown to No.19 MU at St. Athan for storage prior to disposal. It was bought from the RAF by Hawker Siddeley Aviation on December 12th 1968 and placed in store firstly at Hawarden and then at Woodford. In 1973 the machine was bought by the Museum, as having local connections, in the British Aircraft Preservation Council purchase of T. Mk.11s from HSA. XD593 arrived at Newark on March 30th 1973 by road.

General Aircraft ST-12 Monospar VH-UTH

Built 1935, constructor's number 36, light cabin twin, powered by two De Havilland Gipsy Majors of 130 horse power each.

This interesting aircraft (unique in the United Kingdom) was built by General Aircraft Limited at Hanworth, for an Australian customer, as VH-UTH. It served in Australia for over twenty years. In 1961 it was returned to England in an epic flight, taking six weeks.

The story of this eventful 12,000 mile journey can be found in the book 'Two Men in a Flying Machine' by J. Morris. It was intended to keep the Monospar flying, but plans for this failed to materialise and the aircraft languished for a long time, taking a high toll of its external appearance. VH-UTH was acquired by Doug Revell and Mike Hutchinson, becoming one of the founding exhibition pieces. Restoration is currently underway on this historic aircraft, to bring it back to its former glory, though only as a static exhibit.

Gloster Meteor FR. Mk.9 VZ608

Converted 1951 onwards, engine test-bed, powered by two Rolls-Royce Derwent 8 turbojets of 3,500 pounds static thrust each and fitted with one Rolls-Royce RB.108 lift engine.

VZ608 was built by Glosters as a standard FR. Mk.9 photo-reconnaissance fighter at Hucclecote. It served the manufacturers first, being based at nearby Moreton Valance. In March 1951 it was transferred to Rolls-Royce at Hucknall, where the aircraft was to help in the testing of the RB.108 vertical lift jet engine. The RB.108 was originally installed in the fuselage behind the cockpit in the horizontal position. Trials in this way began in 1955. With the third engine in place of the Meteor's main fuel tank, only the underwing tanks could be used for fuel. This limited the aircraft to some thirty minutes flying time. VZ608 was then modified by F. G. Miles Limited at Shoreham to take the RB.108 in the vertical position, and it was given an exact replica of the intake grating from the Short SC.1 VTOL aircraft, the aircraft for which the RB.108 was to be first applied. After the RB.108 had proved itself, VZ608 was not longer needed. It was discovered derelict in February 1970 by the Museum staff, and the aircraft was bought for the collection. VZ608 is still displayed in test-bed guise, although the RB.108 is no longer fitted.

Lee Richards Annular Biplane replica BAPC.20

Built 1964, static film replica.

This curious-looking device was built as a non-flying replica for the film 'Those Magnificent Men in their Flying Machines'. It is based upon an original British pioneering design, which featured a remarkable circular (annular) wing. This replica has been given the British Aircraft Preservation Council identity number 20.

Mignet HM.14 Pou du Ciel 'G-AEOF'

Built in the 1930s, single seat ultra-light, powered by one Scott A.2S 'Flying Squirrel' two cylinder piston engine of 25 horse power.

This 'Flying Flea' was built by W. Miller at Horten Kirby in Kent, before the Second World War at the height of the 'Flea craze'. It was tested there and at Rochester, but did not fly. In 1964 it was bought by R. R. Mitchell and D. Collyer. It is known to have actually flown in 1964/5 - under unofficial conditions. It was later given the spurious marks 'G-AEOF' (having never had a real registration) and given to Newark Air Museum. It has been allocated the British Aircraft Preservation Council identity number 22.

Miles M.14A Magister T9738

Built 1940, constructor's number 2005, basic trainer, powered by one De Havilland Gipsy Major in-line piston engine of 130 horse power.

This aircraft was built at Woodley as a trainer for the RAF, and it served with No.15 and No.24 Elementary Flying Training Schools. After the War it was sold off as G-AKAT to the Puffin Aero Club. (The Magister was known as the Hawk Trainer III in civil guise.) With the Puffin Aero Club it was called 'Katie' (from its registration) and later 'Puffin 2'. It later saw service with the Wolverhampton Flying Club. It was finally allowed to decay at Leicester East. Here it was found in 1964 and moved to Newark. After display at Winthorpe in its civilian colours, G-AKAT was resprayedin the yellow trainer colours of its RAF days, and it again wears the serial T9738.

Percival Proctor II G-AHMP

Built 1940, line number H.170, light cabin monoplane, one De Havilland Gipsy Queen 2.

Built by F. Hills and Sons at Barton as BV631 for the Fleet Air Arm as a communications aircraft, G-AHMP served with No.758 Squadron, based at Worthy Down. It took up its civil identity in 1946. After a brief active career in civil life it was donated to Bristol Grammar School. G-AHMP was in a very sorry state at Filton by March 1963. It was purchased by Mr. G. C. Taylor and stored at his home at Clyst Honiton, near Exeter Airport. He and Lieutenant Commander Blake, who flew the machine when it was BV631, loaned the aircraft to Newark. G-AHMP arrived at Newark in a very bad state, being little more than a centre section.

Percival P.40 Prentice T. Mk.1 VR249 'FA-EL'

Built 1948, basic trainer, powered by one De Havilland Gipsy Queen of 250 horse power.

VR249 was ready for collection by the Royal Air Force at Luton on August 11th 1948 and was delivered to the Royal Air Force College Headquarters Wing at Cranwell five days later. It was given the code letters 'FA-EL' and served as such until going to No.9 Maintenance Unit at Cosford on April 1st 1953 for preparation for use with its next unit. This was to be No.1 Air Signals School at Swanton Morley, with whom it was on charge from June 23rd 1953. VR249 was now coded 'J' and served until August 17th 1957 when it was briefly transferred to No.41 Group before being struck off charge on November 27th, as being sold to Aviation Traders Limited. This was just one of hundreds of Prentices bought by the firm for possible civil use, VR249 was to be one of the few to actually make it into civil use. It was ferried to ATEL's base at Rochford on December 9th 1957. It was converted to a banner and glider tower, and flew for the first time in this guise, registered G-APIY, on July 24th 1958. It was used by ATEL as a demonstrator and appeared at the 1959 Society of British Aircraft Constructor's Display at Farnborough, this time in the guise of a six-seater Prentice. It was eventually sold to Wright and Smith Limited, who based it at Biggin Hill. In 1965 it moved to West Germany, where it flew with Laarbruch Flying Club until 1967, named 'Oracle II'. It was flown into Winthorpe on July 8th 1967 to join the early parts of the Newark collection. It has since been refurbished and now wears the codes and colours from its days at Cranwell.

Percival P.56 Provost T. Mk.1 WV606 'P-B'

Built 1955, advanced trainer, powered by one Alvis Leonides 126 radial piston engine of 550 horse power.

This aircraft, like many Provosts, had only a limited operational career, in this case serving only with No.1 Flying Training School at Syerston, with the code 'P-B'. It was then issued to No.1 School of Technical Training at Halton, where it became one of many Provost airframes used by the School. It was allocated the maintenance airframe number 7622M. In 1971 it was declared surplus and bought by the Museum and moved to Winthorpe. It is displayed in the colours of No.1 FTS, the unit being a local one.

Supermarine Swift FR. Mk.5 WK277 'N'

Built 1955, single seat photo-reconnaissance fighter, powered by one Rolls-Royce Avon 114 of 7,175 pounds static thrust.

WK277 was built at South Marston as a pure fighter F. Mk.4 variant. It was ready for collection by the Royal Air Force on May 26th 1955 and was issued into the care of the Controller (Air) that day. It was issued back to the builders on April 6th 1956 for conversion to FR. Mk.5 status. The conversion completed, it was issued to No.23 Maintenance Unit at Aldergrove on March 7th 1957 and stored until being issued to its first unit on March 3rd 1959. This was with No.2 Squadron at Gutersloh, in West Germany, and WK277 served as 'N' until April 7th 1961. During its time with No.2 Squadron the Swift was involved in two flying accidents, one on July 22nd 1959 and the other on October 3rd 1960. Both of these were Category 3 (repair beyond unit capacity) incidents. WK277 was then retired to No.60 MU at Church Fenton. It was issued to No.2 School of Technical Training at Cosford on May 9th 1961, but was not actually delivered there until November 9th. It was allocated the maintenance serial 7119M. It was bought for the Museum by Norman Pratlett after its disposal from Cosford, and is now on display in its No.2 Squadron markings.

Vickers 668 Varsity T. Mk.1 WF369 'F'

see photo 51

Built 1951, constructor's number 528, multi-purpose crew trainer, powered by two Bristol Hercules 264 radial piston engines of 1,950 horse power each.

WF369 was built at Weybridge and flew for the first time on November 21st 1951. It was delivered to No.201 Advanced Flying School at Swinderby on December 7th, receiving the code 'U'. It moved on to No.2 Air Navigation School at Hullavington from October 18th 1954, becoming 'V' and later 'H'. In 1961 it was to move to Finningley, its last active station. Here it was to give sterling service with the Air Engineers School, later named the Air Engineers and Air Electronics Operators School, with both units it was coded 'K'. It ended its days with Finningley as part of No.6 Flying Training School, this time coded 'F'. It was bought from the Royal Air Force in 1976 by Newark Air Museum's then retiring Chairman, Harold Bradshaw, and flown into Winthorpe on April 1st 1976. This was WF369's 7,612th landing, and was the climax of some 8,646 hours in the air.

<u>Note</u> : The Museum has the following aircraft in store, for possible future display :-

Auster J/1N Alpha G-AMUJ c/no. 2791, remains after crash June 8th 1970.
Thruxton Jackaroo BAPC.21 uncompleted example from Thruxton, possibly to fly again.
Westland Wallace K6038 Remains only from Cranwell.
Supermarine Seafire F. Mk.17 SX336 A2055, with Neville Franklin for rebuild, from a scrapyard in Warrington, also parts from SX300/A646.

NORFOLK & SUFFOLK AVIATION MUSEUM

The Norfolk and Suffolk Aviation Museum saw its beginnings in the Norfolk and Suffolk Aviation Society which was formed by six aviation enthusiasts, who, after looking around, found that there was no organisation catering for their interests in the area. They formed their own Society in December 1972, with an initial membership of 45. The Society encompassed all aspects of aviation. Naturally, preservation of aircraft and items connected with aviation was one of the primary aims. In furtherance of this, the Society obtained Avro Anson G-AWSA from Norwich Airport in January 1974. Soon afterwards, a Spitfire full-size model was donated by the Battle of Britain Museum and both these aircraft were placed in a Wartime Nissen hut near Bungay in Suffolk.

The Society's collection was then moved to Flixton and the Norfolk and Suffolk Aviation Museum was born. However this was only the start and many months were then to pass with preparing the building and exhibits for public display. May 1975 saw the arrival of Meteor WF643 and exactly a year later on May 8th 1976, Wing Commander Kenneth Wallis of autogyro fame, opened the Museum to the public. After a successful first year, the Museum was asked to find new premises and it has now moved a matter of a few hundred yards to the Flixton 'Buck' public house where the small exhibits are displayed in a large brick barn and the aircraft in an enclosure outside.

The Museum is open from May to October, each Sunday from 1000 until 1700 hours. Admission is free, with donations welcome. Visitors should come via Bungay, Suffolk, taking the B1062 to Flixton.

Address : R. W. J. Bullen, Secretary, Norfolk and Suffolk Aviation Museum, 46, Lockhart Road, Ellingham, Bungay, Suffolk, NR35 2HB.

Avro 652A 19 Series 2 G-AWSA

Built late 1940s, general purpose twin, powered by two Armstrong Siddeley Cheetah 15 radial piston engines of 420 horse power each.

G-AWSA was built as an Anson C. Mk.19 general purpose communications aircraft for the Royal Air Force, serial VM349. In 1968 it was retired to No.27 Maintance Unit at Shawbury and put up for disposal. On October 21st 1968 it was registered as G-AWSA to American Jeff Hawke, who intended exporting it to the U.S.A. From November of that year it was stored at Horsham St. Faith. Nothing happened until the August of the following year, when it was given the American registration N5054. However, the aircraft continued to languish at Horsham (Norwich Airport) and fell in dereliction. In January 1974 it was acquired by the Museum and moved into store, later to appear at Flixton for display in the 1976 season. It has since moved to the new site.

Gloster Meteor F. Mk.8 WF643 'X'

Built 1951, single seat fighter, powered by two Rolls-Royce Derwent 8 turbojets of 3,600 pounds static thrust.

WF643 was built by Armstrong Whitworth's at Baginton and was ready for collection by the Royal Air Force June 29th 1951. It was delivered to No.56 Squadron on July 3rd. On May 25th 1952 it was involved in a Category 3 (repair beyond unit capability) accident & was taken on the charge of No.58 Maintenance Unit from Newark for repair on site from May 27th, this work was completed on January 6th 1953. WF643 moved on to No.1 Squadron on March 24th of the following year. Here it was to be inflicted with a more serious accident - Category 4 - beyond repair on site - on July 15th 1954, it was entrusted to No.71 MU at Bicester, but had to go into the charge of Glosters for repair, from September 27th. It was back in the air by May 12th 1955 and was issued to No.8 MU at Little Rissington four days later. From here it was taken on charge with No.611 Squadron only to be hurt again to Category 3 standards on October 21st 1956. No.56 MU despatched a team from Cambridge eight days later to repair it, the team accomplishing their task on November 23rd, and WF643 going back to No.611 Squadron on December 7th. It was retired to No.12 MU at Kirkbride on February 15th 1957. WF643 was declared non-effective stock on September 26th 1957, but was reprieved and re-categorised as effective stock on January 10th 1958. It was then provisionally allocated to the Middle East Air Force on October 9th 1959, and was issued to No.29 Squadron on the

23rd. On September 11th 1963 it was taken on charge with No.103 MU at Seletar in Singapore but re-joined No.29 Squadron, coded 'X' again on October 10th 1963. It was finally retired to No.5 MU just over two years later, on October 22nd 1965. It was declared a non-effective airframe for the second time in its life on September 3rd 1971 and this time the appelation stuck. It was struck off as scrap and taken to Coltishall for use on the dump as a medium for crash-rescue practice. It was acquired in reasonable condition by the Museum and moved to Flixton on May 10th 1975, and has since moved on to the new site at the 'Buck'.

NORTH EAST AIRCRAFT MUSEUM

The North East Vintage and Veteran Aircraft Association began life in 1974 as a small group of vintage aircraft enthusiasts, meeting very informally at Sunderland Aero Club, to exchange views, slides and information on their chosen hobby. At that time the North East was the only area not really covered by an aircraft preservation society.

As the number of people attending the Group's meetings increased rapidly, came the realisation that an informal group was out of the question. Thus NEVVAA was born. At about the same time reports were received that an ex-Fleet Air Arm Dragonfly was rotting in a North Yorkshire scrapyard, and that a Meteor F. Mk.8 was to be broken up at RAF Acklington. After many hurried conversations and much frantic money-raising, it was decided that NEVVAA should go into the aircraft preservation business, and it snowballed from there.

Finding a suitable site to display the embryo collection was the first difficulty. Most avenues explored ended up against a brick wall called 'Money'. However the organisers at Lambton Pleasure Park did show real interest, and were kind enough to provide a site, free of charge. Once the site was sealed the exhibits were moved in.

The Association also undertakes sorties into the 'aerowreckology' field, runs coach trips, holds regular meetings and publishes a journal. At present the Association meets at Newcastle upon Tyne Polytechnic, Ellison Building, room A214 on the second Wednesday of the month, at 1900 hours. Anyone interested is more than welcome.

At the moment the Lambton display are the Dragonfly, the Meteor and the Jet Provost. The Vampire is currently away being refurbished but may re-join the display. Other aircraft are held in store, as detailed below.

Lambton Pleasure Park is well sign-posted at Chester-le-Street, near Sunderland. General entrance to the pleasure park, which includes a wild life reserve, must be made to view the collection. Details of current admission fees to the Pleasure Park can be had from :- Lambton Pleasure Park, Lambton Hall, Chester-le-Street, Durham.

NEVVAA are at present re-organising themselves and hope to change their name with the establishment of Limited Liability Company status. At the same time, plans are underway to establish a Museum in the Association's own right at Sunderland Airport - Usworth. It is hoped that this plan will come to fruition in 1977, in which case the aircraft at Lambton will move to the new site. Up to date information on this, and any queries can be had by writing to the following address, please enclose a stamped addressed envelope.

Address : NEVVAA, 35, Grosvenor Place, Jesmond, Newcastle upon Tyne, NE2 2RD.

British Aircraft Corporation BAC.145 Jet Provost T. Mk.5 XW331

Built 1970, basic trainer, powered by one Rolls-Royce Viper 522 turbojet.

This aircraft was built at Warton and served with the Royal Air Force only briefly before coming to grief in a bad landing at Leeming in 1975. It had been in service with No.3 Flying Training School, with the code '46', and had seen a season of performances with that unit's display team, the 'Gemini Pair'. After being stripped of all useful components the hulk was sold off to a Yorkshire scrap merchant. It was quickly bought up by the Museum and moved to Lambton Pleasure Park where its restoration has continued to date. As parts of the T. Mk.5 are difficult to come by, except direct from the manufacturer at huge cost, the restoration is relying on the Jet Provost T. Mk.4 to complete the aircraft, this will make it a unique hybrid. The port wing is from XP618 (also ex No.3 FTS as '50') and the starboard wing is from XP674 (again ex No.3 FTS as '65') both of these components were obtained from the fire dump at Ouston. Other parts are also in use, including the rear end of XP661, maintenance number 7819M, previously with No.6 FTS, coded '44', and previously derelict at Linton-on-Ouse. This piece is in use as a donations collection box in front of the exhibit.

De Havilland DH.115 Vampire T. Mk.11 WZ518 '37'

Built 1953, constructor's number 15105, advanced trainer, powered by one De Havilland Goblin.

WZ518 was built at Hawarden, and was ready for collection by the Royal Air Force in March of 1953. It first served with No.14 Squadron as a continuation trainer and 'hack'. It was to continue the latter role with its next unit, the 2nd Allied Tactical Air Force, Germany, Communications Flight, followed by a spell with the Oldenburg Station Flight. It returned to the United Kingdom and became part of No.5 Flying Training School at Oakington, receiving the code '37'. It was sold to Hawker Siddeley Aviation on October 30th 1967 and placed into store at Hawarden, later moving on to Woodford in 1971/2. After failure to be sold to another customer it was sold for a token payment to the Stockport Aircraft Preservation Society, via the British Aircraft Preservation Council. It was delivered by road to Pomona Dock in Manchester in 1974 and erected next to a Comet 2 set up there as a restaurant. HSA had mixed the fuselage and booms with the wings of WZ608, making two hybrid aircraft with NEVVAA. WZ518 was moved to SAPS's store at Handforth but was soon disposed of by that group and sold to NEVVAA. It was delivered to Lambton Pleasure Park on October 11th 1975. It has since been restored using parts from XE894 (acquired from British Airways at Heathrow) at Usworth, and may re-join the exhibits at Lambton.

Gloster Meteor F. Mk.8 WL181 see photo 49

Built 1954, single seat fighter, powered by two Rolls-Royce Derwent 8 turbojets of 3,600 pounds static thrust each.

WL181 was ready for collection by the Royal Air Force on May 26th 1954 and was delivered to No.20 Maintenance Unit at Aston Down two days later. It joined the Station Flight of Tangmere on July 16th for a brief spell before being taken on charge with No.34 Squadron om August 6th. It was flown to No.29 MU at High Ercall on December 7th 1955, followed by No.12 MU Kirkbride from February 11th 1957, and No.33 MU at Lyneham from September 17th 1959. After this period of rest it was delivered to the College of Air Warfare at Manby on March 28th 1960. Here it was to wear the code 'G' and also become part of the unit's aerobatic team - the 'MACAWS'. It completed its operational flying with the Advanced Jet Refresher Squadron at Strubby, before being retired to No.5 MU at Kemble on May 3rd 1965. It was declared a non effective airframe on September 10th 1971 and struck off charge on November 3rd. It was moved to Acklington in February 1972 to act as an instructional airframe for No.22 Squadron to provide winching practice for the unit's air sea rescue helicopters. By the end of 1974 Acklington had been closed and WL181 was threatened with the axe. It was bought by NEVVAA on March 18th 1975 for £200 and moved to Lambton by road on May 3rd 1975.

Westland WS-51 Dragonfly HR. Mk.5 WG724 '932'

Built 1952, line number WA/H/55, general purpose naval helicopter, powered by one Alvis Leonides 521 radial piston engine of 500 horse power.

WG724 first flew on October 29th 1952 at Yeovil and was delivered to the Aircraft Holding Unit at Gosport on November 17th and was stored. It joined the Station Flight at Ford from February 1956 until March 1958. In 1958 it was moved to the Royal Naval Aircraft Yard at Donibristle for conversion from its HR. Mk.3 standard to HR. Mk.5 status. It was then in use with Airwork at Hurn from September 1958 until February 1960. At this time it suffered a Category 4 accident and was taken to the RNAY at Fleetlands for repair in March. This was not completed until September 1961 and WG724 was then stored until February 1962 when it joined the Station Flight at Lossiemouth, coded '932/LM', serving until May 1963, when it returned to Fleetlands. It was struck off charge on October 3rd 1965 and was sold to the Anglo-Diesel Company of London and moved to Blackbushe the following month. It was to become a familiar part of the landscape at Blackbushe, as part of a gaggle of Dragonflies held in external storage there by the company. It moved in March 1968 to Mr. Else's scrapyard at Wisbech before moving, in August 1970, to the Forway Garage at Moor Monkton and later into Hugh's scrapyard in that town. It was from here that NEVVAA purchased the craft for £150 in March 1975 and moved it to Lambton Pleasure Park.

Note : NEVVAA have the following aircraft in store, which may be displayed in the future :-

Bensen Gyroglider Built by Mr. F. Brown, in Stanley, using the rotor from BAPC.95 below.
Brown Helicopter BAPC.96 Built by Mr. Brown to his own design, twin rotor helicopter powered by a converted McCulloch engine.
Gizmer Autogyro BAPC.95 Built by Mr. Brown to American plans. Damaged at Middleton St. George in the early 1960s in a running-up accident. Rotor donated to the Bensen.
Luton LA-4 Minor BAPC.97 Built by Mr. Brown 1937-1944. First with a Sumet 'V' engine, then an original design five cylinder radial, A.B.C. Scorpion and finally a JAP J-99. Later converted to have folding wings. Despite many attempts did not fly.

NOSTELL AVIATION MUSEUM

The Museum was formed in 1976 when the South Yorkshire Aviation Society entered into association with Lord St. Oswald's estate at Nostell and No.127 Squadron, Air Training Corps from Wakefield. It contains the Society's collection of aviation items, including the Dixon Boot collection of flying clothing. The Society was also able to obtain a number of aircraft and engines on loan for display. The Museum was inaugurated by Marshal of the Royal Air Force Lord Elworthy on 11th July 1976.

The Museum is open at weekends from April to October between 1400 and 1800 hours. Admission is adults 15p and children 10p.

Addresses : Nostell Aviation Museum, The Curator, Nostell Priory, Wakefield WF4 1QF.
South Yorkshire Aviation Society, 15, Troon Road, Hatfield, Doncaster, DN7 6SJ.

Bristol B.171 Sycamore HR. Mk.14 XG504

Built 1955, constructor's number 13265, general purpose helicopter, powered by one Alvis Leonides 73 radial piston engine of 520 horse power.

XG504 was completed on March 31st 1955 and was first issued to the Aden Communications Squadron on July 9th. It was transferred to the Khormaksar Station Flight on June 1st 1957 and then to that station's SAR Flight, before going to No.131 Maintenance Unit for shipping back to the United Kingdom on March 21st 1964. It was taken on charge at No.15 MU at Wroughton on April 3rd. On August 12th 1965 it was moved to Royal Naval Aircraft Yard Fleetlands for reconditioning before joining the Central Flying School at Ternhill on December 20th 1965, coded 'S-S'. It was stored at No.15 MU from May 26th 1966, moving on to No.5 MU at Kemble on December 28th 1967. It was issued to the Metropolitan Communications Squadron at Northolt on April 18th 1968, this unit changing its name to No.32 Squadron on February 13th 1969. It was retired and bought by Mr. J. McArdle on August 14th 1972 and moved to Rotherham. It was loaned to Nostell in 1976.

Mignet HM.14 Pou du Ciel 'G-AEKR'

see photo 78

1970 reproduction static replica, fitted with an original Anzani piston engine of 25 h.p.

This aircraft is a replica (using only the engine of the original) built in 1970 by the staff of RAF Finningley to replace the original which was destroyed in a hangar fire at the base in 1970. Indeed the 'Flea' was used to start the fire by the vandals who had broken in! After display at Finningley the new aircraft moved to the Motor Museum at Crowle for the 1975 season and was finally loaned to the Museum by Mr. B. Claybourne in 1976. The original G-AEKR, constructor's number CAC.1, was built in 1936 by W. Claybourne and Company Limited and carried 'This is the Flea that Claybourne's Built' on its rudder. It was flown with a fair degree of success from Doncaster. It was cancelled from the register in July 1938 and stored until presented on loan to Finningley and displayed as part of the Station Museum until the fire.

Mignet HM.14 Pou du Ciel 'G-AFFI'

Built 1973, constructor's number YA-1, single seat ultra-light, powered by one Scott A.2S 'Flying Squirrel' two cylinder piston engine of 25 horse power, aircraft taxiable only.

'G-AFFI' (the registration was not originally allocated, and was chosen as it fits in with many original 'Fleas') was started on August 12th 1973 at Rawdon near Leeds, by Dave Allen, Peter Barber and Steve Langfield, these traded as Yeadon Aeroplanes. The aircraft has been allocated the British Aircraft Preservation Council identification number 76. It was finished in 1975 and displayed at the 1975 and 1976 Popular Flying Association's Fly-ins at Sywell. It was delivered on loan to Nostell on August 30th 1976.

Saunders-Roe Skeeter AOP. Mk.12 XL764

Built 1958, constructor's number S2/5076, aerial observation post, powered by one De Havilland Gipsy Major 215 in-line piston engine of 215 horse power.

Built at Eastleigh, XL764 was ready for collection by the Army Air Corps on October 9th 1958. By August 11th 1959 it had suffered an accident and had to be repaired on site by the manufacturers, this work was finished on October 19th. On January 9th 1962 it was again the victim of a flying accident, this one given as Category 3 (beyond repair by the local unit) and it was moved to the care of Westlands (who had absorbed SARO) on April 4th. The debending was completed on January 17th 1964. On the 20th it was transferred to the Ministry of Aviation and displayed at the Society of British Aircraft Constructor's Display at Farnborough in September. On March 13th 1967 it was allocated the maintenance airframe number 7940M at Arborfield. It was moved to Middle Wallop in July 1975 and was bought by Mr. J. McArdle of Rotherham. It was loaned to the Museum in 1976.

Note : The following aircraft are stored at Misson for the Museum :-

Gloster Meteor NF. Mk.14 WS739 7915M previously on the gate at Church Fenton. SYAS also have the nose section of an anonymous Meteor NF. Mk.11.
Bristol B.171 Sycamore HR. Mk.14 XG506 ex Central Flying School, coded 'S-P', acquired from No.1 School of Technical Training at Halton in 1975 as 7852M.

ROYAL AIR FORCE MUSEUM

The Royal Air Force Museum is the only British national museum devoted solely to the history of aviation, and to the complete story of one service (including its predecessors) present and future.

That it should be situated at Hendon is indeed appropriate, for the locale is the cradle - if not quite the birthplace - of British aviation. As far back as 1862 Henry Coxwell and James Glaisher were flying a balloon, the Mammoth, from the nearby heights of Mill Hill to far distant Biggleswade. Claude Grahame-White brought his flying training school to Hendon and Louis Bleriot soon followed him to the airfield and established his school. In 1914 the airfield was commandeered for war service and the rich history, that included the Air Pageants which were to make Hendon synonymous with aviation throughout the world, was under way.

Today, although the airfield is now built-over, much of the old Hendon remains; it still houses an active Royal Air Force station wherein buildings of the utmost modernity stand side by side with some of Grahame-White's original structures; and, permanently, it houses a national museum that starts with the military balloonists of the 1870s and details the profuse pageantry of powered flight from the days of the Royal Flying Corps and the Royal Naval Air Service up to the current and future activites of the Royal Air Force.

In addition to its displays the Museum contains a restaurant, a cinema-lecture hall, and an important research centre containing a unique collection of aviation records covering MSS, books, documents, plans, films, photographs, and other relevant material covering the span of British aviation history.

The Museum has several separate displays, the main hall carries the bulk of the aircraft exhibits; the Sidney Camm Memorial is situated in the midst of this and is a collection of machines designed by Hawker's prolific designer; there is an external display of larger aircraft (occasionally supplemented by theme displays associated with the Dermot Boyle Wing); the Galleries which contain many displays and some aircraft sections; and the Dermot Boyle Wing which is described below.

The Royal Air Force Museum was officially opened by Her Majesty Queen Elizabeth II on November 15th 1972. An extension, The Dermot Boyle Wing, was officially opened by His Royal Highness The Prince of Wales, in November 1975.

The Dermot Boyle Wing is designed to stage special exhibitions to depict aspects of aviation history hitherto untold in three-dimensional form, drawing upon items from major collections throughout the world as well as upon the Museum's own extensive holdings. The first such exhibition was 'Wings of the Eagle', opened by Prince Charles and depicting the history of German aviation. In February 1977 a display to mark the Queen's Silver Jubilee, entitled 'Kings and Queens and Flying Machines' was opened. This is planned to be followed by an exhibition on the history of French aviation.

The Royal Air Force Museum has a huge stockpile of display items and whole aircraft, which are centred on the restoration and storage facilities of Cardington and Henlow. Other reserve collection aircraft are held at operational Royal Air Force stations in store, or under restoration to display condition, there are also three large 'satellite' collections of reserve aircraft at Cosford, Finningley and St. Athan. The former is now open full time to the public and can be found in this work in its own right, under the heading Cosford Aerospace Museum. Finningley and St. Athan are not generally open to the public, although their aircraft are available for inspection by prior arrangement with the Officer Commanding, and are open to the public every Battle of Britain Day. The reserve collection aircraft are mentioned after the main RAF Museum listing. In addition, the aircraft of the Battle of Britain Memorial Flight are also technical part of the RAF Museum, these aircraft can be seen at many air displays up and down the country. These, too, are listed at the end of the main text on the RAF Museum.

The Museum is conveniently situated to be reached from the Colindale station of the Northern Line of the London Ungerground System; it is also very close indeed to the Hendon exit of the M1 motorway. The Museum is open from 1000 to 1800 hours on weekdays, and from 1400 to 1600 hours on Sundays. Admission is free, but it is possible that a small charge may be made from time to time for admission to special exhibitions, especially those held in the Dermot Boyle Wing.

Address : The Royal Air Force Museum, Hendon, London NW9 5LL.
Telephone : 01 205 2266 - four lines.

Avro 504K E449

Reconstruction, completed in 1966, basic trainer, powered by one Clerget rotary piston engine of 130 horse power.

This aircraft is a hybrid and part of the collection of aircraft, and other forms of transport, built up by R. G. J. Nash, and known as the Nash Collection. In December 1953 this, and most of the collection was dispersed, this aircraft being bought by the Royal Aeronautical Society, it was handed over to the Museum in March 1963. This Avro was rebuilt to exhibition standard, as E449, and was displayed at the Royal Review of the Royal Air Force 1968. E449 was then stored at Henlow, until mounted inside the Museum in 1972. The fuselage of this aircraft comes from Avro 504K G-EBJE which was registered in January 1926 to the Southern Counties Aviation Company, passing on to F. G. Miles in August 1926. It was withdrawn from use officially in September 1934. The wings and other parts of the aircraft (although much is of contemporary build) come from E449 which was built as an Avro 504K by Harland and Wolfe at Belfast. It was registered to the Aircraft Disposal Company at Croydon as the prototype Avro 548A G-EBKN in April 1925. The 548A was a re-engined 504K airframe, with a 80 h.p. Renault. The remains of both aircraft were bought up by the Nash Collection in the 1930s.

Avro 683 Lancaster I R5868 'PO-S'

Built 1942, long range heavy bomber, powered by four Rolls-Royce Merlin XX piston engines of 1,480 horse power each.

'S-Sugar' is probably the most crowd-pulling aircraft at present housed in the Royal Air Force Museum. The history that follows here is but a mere glimpse at its past, the reader is recommended to read the RAF Museum's own book on the subject 'A Very Special Lancaster', by F. E. Dymond, which is available at the Museum bookshop. R5868 was built by Metropolitan-Vickers at Trafford Park and was assembled at Woodford and awaiting collection by the RAF on June 20th 1942. It was delivered to No.83 Squadron at Scampton nine days later, and took up the unit's code letters 'OL-' and the individual letter 'Q'. Its first operational sortie came on the night of July 8th/9th 1942 with Squadron Leader R. L. Hilton D.F.C. at the controls when R5868 dropped 1,260 four pound incendiary bombs onto Wilhelmshaven, this trip taking some four hours thirteen minutes to complete. No.83 moved to Wyton on August 15th, to take up a new role with the Pathfinder Force. On the night of September 14th/15th R5868 was again over Wilhelmshaven, again with Squadron Leader Hilton, and dropped eight 250 pound incendiaries, but, more importantly, six sticks of four marker flares. On the return flight the wireless operator was wounded by fire from another four-engined twin fin aircraft overtaken over the sea! R5868 flew its last operational sorties with No.83 Squadron all the way to Milan on the night of August 14th/15th 1943, the 'Lanc' clocking eight hours seven minutes in the air. Here it dropped four yellow target indicators, one 4,000 pound bomb and three 500 pound bombs. R5868 was then transferred to No.467 Squadron Royal Australian Air Force, based at Bottesford. Here it was taken on charge with 'B' Flight of the Squadron, and given the unit code letters 'PO-' and the immortal individual code 'S', which turned 'Queenie' into 'Sugar'. 'Sugar' made its first sortie with the new unit (its 78th operational sortie) under the command of Pilot Officer A. M. Finch to Hanover on the night of 27th/28th September 1943. In October 'Sugar' was briefly lent to No.207 Squadron at Langar to make up for an aircraft shortage with the unit. A note in the official records puts 'Sugar' as fit only for operational conversion training at this point. On November 11th 1943 the Squadron moved to Waddington. With Flying Officer J. A. Colpus in control 'Sugar' faced a serious accident over Berlin on the night of November 26th/27th. Bomb load that night was one 4,000 pounder, 56 thirty pounders and 1,050 four pound incendiaries. 'Sugar' was in collision with another Lancaster just after bombing the target. The huge craft went into a severe dive to port, but by the use of rudder, aileron and engines the aircraft managed to maintain height and landed at Tholthorpe. On completion of about 88 operational sorties in March 1944 'Sugar' (who had been receiving a bomb on the nose for every operational raid) was adorned with the words of Herman Goering, Marshall of the Luftwaffe - "No enemy plane will fly over the Reich territory". Shortly after this 'Sugar' bombed Aulnoye on the night of March 25th/26th with the port outer engine inoperative, the weary bomber only just making it into Tangmere. On the night of May 11th/12th 'Sugar' managed to successfully out-fly a pair of Junkers Ju 88 night-fighters by forestalling them, although it was ordered not to bomb that night, possibly because of the attack. After much campaigning 'Sugar' was dismantled and moved by road to Woodford on August 3rd 1944 for a well-earned refurbish, this work being completed on November 17th, 'Sugar' re-joining No.467 Squadron on December 3rd. 'Sugar' continued to give excellent service. On a daylight raid on April 6th 1945 it again did not drop its bomb-load. The target was Ijmuiden and the load was a devastating fourteen thousand pound bombs. The Allied forces had occupied the town by the time the bombers had arrived! 'Sugar' made her last sortie, *its one hundred and thirty seventh*, to Flensburg on April 23rd 1945 with a load of eight one thousand pounders and six five hundred pounders, in the hands of Flying Officer L. W. Baker, no attack was eventually made however. On August 23rd 1945 R5868 was officially transferred to No.15 Maintenance Unit at Wroughton. It was the end of an amazing operational career, 'Sugar' had flown some 795 hours 25 minutes on operational flying and dropped about 466 tons of bombs operationally. R5868 was struck off charge on March 16th 1956 and moved to Scampton in 1959 for service as a gate guardian, receiving the airframe number 7325M, and was put back into No.83 Squadron's colours as that unit was resident at Scampton, flying Vulcans. On November 24th 1970 R5868 was dismantled and removed from the gate by a team from No.71 MU who were to refurbish it for the RAF Museum at Bicester. It moved to Hendon on March 12th 1972 and was painted back in No.467 Squadron colours.

Blackburn B.101 Beverley C. Mk.1 XH124

Built 1957, long range large capacity transport, powered by four Bristol Centaurus radial piston engines of 2,850 horse power each.

XH124 first took to the air on April 1st 1957, from Brough, and was officially ready for collection by the Royal Air Force on April 30th. It was flown over to No.27 Maintenance Unit at Shawbury on May 1st and prepared for service. It was delivered to No.30 Squadron at Dishforth on July 2nd, becoming 'G' with that unit. XH124 was detached to Aden from November 15th through to February 27th 1958 with the unit, and this was to be the craft's first association with the province. From May 21st 1959 XH124 was loaned to No.84 Squadron and was based in Aden at Khormaksar until July 31st 1959. It returned to Dishforth and No.30 Squadron and was detached to Eastleigh, Nairobi, from October 21st 1959 and was to involve itself in relief work in the Belgian Congo, and, in July 1961 played a part in the Kuwait emergency, using Bahrain as a base. In March 1962 it flew home for a re-skinning of its wings. On December 16th 1964 XH124 landed at St. Athan from Muharraq for an overhaul and camouflaging by No.32 MU there. After this treatment it was handed over to a new unit, No.242 Operational Conversion Unit at Thorney Island on March 15th 1965 to take up the role of training 'Bev' crews. It last served with No.47 Squadron briefly from Abingdon from March 2nd 1967 until April 4th 1967, when it was flown into Shawbury for storage. It was prepared for display at the 50th Anniversary of the RAF celebrations at Abingdon. From here it made its last flight into Hendon on June 19th 1968 (five days after the Abingdon review), gaining the distinction of being the last fixed wing aircraft to land at Hendon. With this XH124 had completed some 4,478 hours in the air. It was allocated the maintenance airframe number 8025M on June 27th.

Bleriot Type XI No.164

Built circa 1909, single seat light plane, powered by one Anzani two cylinder piston engine.

This aircraft was part of the Nash Collection, and is now technically on charge with the Royal Aeronautical Society and on loan to the Museum. Little is known of its history, its identity being its constructor's number, and not a military serial. Details welcomed.

Boulton Paul P.82 Defiant I N1671

Built 1940, two seat turret-equipped fighter, powered by one Rolls-Royce Merlin.

N1671, destined to be the only Defiant to survive intact into the 1970s, was taken on charge by the Royal Air Force on August 7th 1940 at No.6 Maintenance Unit at Brize Norton. It was delivered to No.307 (Polish) Squadron at Kirton-in-Lindsey on September 17th, and was coded 'EW-D'. The Squadron was dispersed to Caistor on October 2nd, returning to its original base on the 27th. On November 7th the unit moved to Jurby, for operations that basically entailed shipping protection. On January 14th 1941, N1671 was again back at No.6 MU, but it returned to No.307 Squadron, now based at Squires Gate, on the 27th. On March 26th the unit migrated to Colerne, moving on to Exeter on April 26th, all the time being involved in coastal sweeps. On June 9th 1941, with Sargent Bilau in command the Defiant was dispersed to Church Stanton, the ten minute flight ended in a Category 3 landing

55 *Hawker Hart J9941 at Farnborough 13/9/64, then with HSA, now with the Royal Air Force Museum*
Stuart Howe

56 *Gladiator K8042 camouflaged but without roundels, awaiting display in the RAF Museum*
via Les Hunt

57 *Mosquito TW117 at Farnborough 1963* *P. H. Butler*

58 *Lysander R9125 at the Royal Review 1968 R.A. Scholefield*

59 *Hoverfly KK995 at Henlow, May 1970* *Stuart Howe*

60 *Wellington MF628 in store at Biggin Hill*

61 *SR.53 XD145 also at Henlow May 1970* *Stuart Howe*

62 *Bright red Avro 707 WZ736 at Finningley* *S. G. Jones*

accident, and it was taken on charge with Reid and Sigrist on June 14th for repair. Fixed, N1671 was flown to No.46 MU at Lossiemouth by October 10th 1941. It was delivered to No. 153 Squadron at Ballyhalbert twenty days later. On June 22nd 1942 it joined No.285 (Anti-Aircraft Co-operation) Squadron at Wrexham, moving with the unit to Honiley on October 29th. This unit maintained detachments at Wrexham and Woodvale, and later at High Ercall, and N1671 may well have operated from these fields. It was in the repair workshops at Honiley from February 23rd 1943 until April 21st. On May 16th it was on charge with No.10 MU at Hullavington, moving to No.52 MU, Cardiff, on September 8th 1944. It was struck off charge on June 21st 1947 and was to spend a long time in store, being at No.2 MU at Stanmore Park by September 1955 and at Fulbeck by January 1960. In December 1967 it was moved to St. Athan for refurbishment by No.19 MU for exhibition at the Abingdon display in June 1968. The following month it was moved to the Station Museum at Finningley until being moved by road to Hendon on April 1st 1971.

Bristol 156 Beaufighter TF. Mk.10 RD253

Built 1944, two seat torpedo carrying fighter, powered by two Bristol Hercules XVIIs.

RD253 was built by Bristol's at Oldmixon, and delivered to No.19 Maintenance Unit at St. Athan in November 1944. On March 7th 1945 it was taken on charge at Pershore for preparation for sale to the Portuguese Navy. It was flown out to Portella from Pershore ten days later, by No.1 Ferry Unit. The 'Beau' was given the Portuguese serial BF-13 and was used for a short period on maritime protection work. It was later relegated to the role of ground instructional aiframe, and was eventually given to the Lisbon Technical Institute about 1950. Here it was to stay outside until being presented to the Royal Air Force in 1965. RD253 was moved into the care of No.71 MU at Bicester, who already had RD867 from Malta. The latter was rebuilt using parts from the former. The remains of RD253 were given to No.4 School cf Technical Training at St. Athan who managed to rebuild it totally. This aircraft was chosen to be put on display at the Museum and it arrived by road at Hendon on March 15th 1971.

Bristol 192 Belvedere HC. Mk.1 XG474 'O'

Built 1962, heavy transport twin-rotor helicopter, powered by two Napier Gazelle 100 turboshafts of 1,650 horse power each.

XG474 was also built at Oldmixon, and was ready for collection by the Royal Air Force on March 14th 1962. In May it was shipped to Seletar, Singapore, and joined No.66 Squadron as 'B' on June 8th. It was shipped to the Middle East and taken on charge at Khormaksar, Aden, on May 15th 1964, taking up charge with No.26 Sqn, still coded 'B' on the 23rd. Here it was used in operations against the terrorists in the Radfan area. It was shipped back to Seletar on January 26th 1965 and was back on charge with No.66 Squadron on December 7th. It was uncoded during 1966, but had become 'O' by 1967. No.66 Squadron disbanded in March 1969 and XG474 made its last flight on the 20th, with a total of 1,675 hours in the air. It was earmarked for the RAF Museum and was shipped back on board the aircraft carrier H.M.S. 'Albion', arriving at Portsmouth on August 1st 1969. It moved to Fleetlands and then on to Henlow for storage. It moved to Hendon early in 1971.

Caudron G.III '3066'

Built circa 1916, constructor's number 7487 (unconfirmed), two seat military aircraft, powered by one Anzani radial piston engine of 90 horse power driving a pusher propellor.

This aircraft was finished at Henlow to represent G.III (N)3066 of the Royal Naval Air Service Flying School at Vendome in 1917. In actual fact this aircraft was registered in Belgium on July 21st 1921 as O-BELA, changing to OO-ELA in 1929. It was bought from J. E. Ledure by Ken Waller and flown Brussels-Ostend-Lympne-Brooklands by Mr. Waller on May 20th 1936. It was displayed at the Hendon Air Display on June 27th 1936. The aircraft was allocated the civil registration G-AETA, but this was not taken up. The Caudron became part of the Nash Collection , and subsequently entrusted to the Royal Aeronautical Society in November 1953. It was handed over to the Royal Air Force Museum, along with the rest of the collection, on permanent loan in 1964. The Caudron was taken to Stradishall and its overhaul was initiated. It was moved to Henlow in 1966 and finished as above, being installed at Hendon in 1972.

Cayley Glider replica BAPC.89

Modern reconstruction, single seat glider.

This replica was built by Southdown Aero Services, under the direction of J. S. Sproule for Anglia T.V. It was presented to the Museum in 1976.

De Havilland DH.82A Tiger Moth II T6296

Built 1941, basic trainer, powered by one De Havilland Gipsy Major in-line piston engine of 130 horse power.

This aircraft was kindly donated to the Museum by the Fleet Air Arm when the Museum's own Tiger, NL985 (parts of which remain at Henlow) was burnt by vandals. T6296 was built by Morris Motors and was first issued to No.1 Elementary Flying Training School at Hatfield on June 13th 1941, going to No.39 Maintenance Unit at Colerne for storage on December 7th. It was issued to No.7 EFTS at Desford on February 17th 1942 until again going into storage on September 23rd 1946, this time at No.5 MU, Kemble. On December 11th 1946 it was transferred to the Fleet Air Arm and was delivered to Royal Naval Air Station Stretton, later going to the Royal Naval Engineering College at Manadon, and then the Britannia Royal Naval College at Roborough, ending its days with the Station Flight at Yeovilton, until being donated to the Royal Air Force Museum on May 9th 1972. It is worth noting here that on June 14th 1951 G-ALAD was registered to the Wiltshire School of Flying as 'ex T6296' - this aircraft was sold in New Zealand in February 1952, becoming ZK-BAD. To this date the real identity remains obscure but it is possible that it was the real T6296 and that the one flying as such until 1972 was another aircraft.

De Havilland DH.98 Mosquito T. Mk.III TW117

see photo 57

Built 1946, advanced trainer, powered by two Rolls-Royce Merlin 21 piston engines.

TW117 was built at Leavesden and was ready for collection by the Royal Air Force on May 15th 1946. It was delivered to No.15 Maintenance Unit at Wroughton fifteen days later. It was first issued to the Armament Practice Station at Acklington on July 22nd 1947 serving there until 1949. On October 18th 1949 it was issued to the Station Flight of Linton-on-Ouse, and loaned by the unit from June 1950 to the based Hornet Conversion Flight, with whom it gained the codes 'MS-A'. From the HCF it moved to No.264 Squadron, coded 'PS-Z' and then to No.58 Squadron at Benson on September 25th 1953. It was retired to No.48 MU Hawarden on April 15th 1954 and stored, later moving to No.5 MU at Kemble from September 28th 1956, back to No.48 MU from November 28th 1956 and finally to No.27 MU at Shawbury from July 19th 1958. It was brought out of mothballs and issued to No.3 Civilian Anti-Aircraft Co-operation Unit at Exeter on March 2nd 1960, with the code 'Z' and a coat of camouflage. TW117 was struck off charge at Henlow on March 31st 1963 on its retirement from life with No.3 CAACU, and was allocated the maintenance airframe number 7805M on May 24th 1963. It was put on display inside Hendon in 1972 in silver scheme.

De Havilland Canada DHC-1 Chipmunk T. Mk.10 WP912

Built 1950s, basic trainer, powered by one De Havilland Gipsy Major 8 of 145 horse power.

This aircraft is the centre-piece of the new exhibition 'Kings and Queens and Flying Machines' in the Dermot Boyle Wing for the first half of 1977, to celebrate the Silver Jubilee of the Queen. It was on this aircraft that His Royal Highness The Duke of Edinburgh earned his 'Wings'. It was in use with the Manchester University Air Squadron, based at Woodvale, but was retired in 1973 and moved for a short spell to Newton, before being flown into No.5 Maintenance Unit at Kemble for storage and disposal on April 31st 1974. Here its significance was realised, and it was not sold off to civil operators, like many other ex RAF Chipmunks. It was overhauled and delivered by road to Hendon on December 14th 1976. 'Chippie' WP912 has 6,450 flying hours to its credit.

English Electric Canberra PR. Mk.3 WE139

Built 1953, constructor's number 71110, three seat reconnaissance aircraft, powered by two Rolls-Royce Avon 101 turbojets of 6,500 pounds static thrust each.

WE139 was awaiting collection by the Royal Air Force on January 30th 1953. It was delivered to the Handling Squadron at Manby on February 3rd, staying here only until May 21st when it moved to the Bomber Command Acceptance and Modification Unit, based at Binbrook. It returned to the makers on June 12th 1953 for modifications, being delivered to No.540 Squadron at Wyton on August 10th 1953. It was entered in the London to New Zealand Air Race of October 1953. The race was run to mark the centenary celebrations of the Province of Cantebury , and offered £10,000 for each of the two sections and the Harewood Gold Cup for the winner of the speed section. There were two sections, a speed dash from Heathrow to Harewood, and a handicap event for other aircraft. WE139 was flown by Flight Lieutenant R. L. E. Burton A.F.C. and Flight Lieutenant D. H. Gannon D.F.C. and took off on October 8th 1953 arriving at Harwood 23 hours 51 minutes later. With several withdrawals the speed section became an all Canberra race, with three from the

RAF and two from the Royal Australian Air Force competing. WE139 was declared the winner of the speed section. After the race publicity had died down, WE139 was put back into active service with No.69 Squadron at Gutersloh, West Germany, from October 1st 1954. The unit moved to Laarbruch in December and on to Luqa, Malta, on April 1st 1958. On July 1st it was transferred to No.39 Squadron, still at Luqa, and was given the code 'C'. Seventeen days later it was flown to No.39 Maintenance Unit at Colerne for a special fit of equipment. It was back on strength at Luqa on September 9th. It returned to English Electric for modifications on May 5th 1960, returning to the unit on October 25th. From December 19th 1962 it was on charge with No.231 Operational Conversion Unit at Bassingbourn. It was flown into Henlow on April 23rd 1969, this being its last flight, WE139 having amassed a total of 4,260 hours in the air. It was moved by road to Hendon on January 14th 1971, and was completed re-assembled on March 25th 1971.

English Electric P.1B XA847

Built 1957, constructor's number 95004, single seat development aircraft, powered by two Rolls-Royce Avon 200 turbojets of 14,430 pounds static thrust each.

XA847 first flew from Warton on April 4th 1957 and was used until 1959 by the makers on trials, leading up to the launch of the true Lightning series. XA847 was the first aircraft of British origin to fly over Mach 2 in level flight, it achieving this piloted by Roly Beaumont on November 25th 1958. It was delivered to the Aeroplane and Armament Experimental Establishment at Boscombe Down in 1959 and was used for numerous trials. It gained the extended fin that it still wears in May 1963. Also in 1963 it was used for rocket trials and was given the leading edge extension of later Lightnings. It was flown to the Royal Aircraft Establishment at Farnborough on April 21st 1966 and used for gravel arresting trials from August, until September 1968. It was then presented to the RAF Museum and removed to Henlow, appearing, refurbished at Hendon in 1972.

Gloster Gladiator II K8042

see photo 56

Built 1937, constructor's number 40468, single seat fighter, powered by one Bristol Mercury radial piston engine of 840 horse power.

K8042 was first used for makers trials, and then gravitated to the Aeroplane and Armament Experimental Establishment at Boscombe Down for acceptance trials with the Royal Air Force. It entered true service with No.5 (Pilots) Advanced Flying Unit, followed by No.61 Operational Training Unit, until being struck off charge on March 16th 1948. It was stored at Biggin Hill for much of the post war period, and was subjected at one time to an inaccurate camouflage paint scheme. It was refurbished and painted up in No.87 Squadron colours and moved into Hendon in 1972.

Gloster Gladiator II N5628

Built 1939, constructor's number unknown, other data as above.

This aircraft is in the exact state that it was found in when pulled out of the icy waters of Lake Lesjakog in Norway in 1970. It is on display at Hendon in this condition to emphasise the work involved in restoring an aircraft to display condition. The tail section and wings etc. are in store at Cardington. N5628 was with No.263 Squadron when it was shipped out to Norway on board the aircraft carrier H.M.S. 'Glorious' on April 22nd 1940 to take part in the defence of Norway. The Gladiators of No.263 were to operate from the frozen lake, in the absence of airfields in Norway. The Gladiators first landed there on April 25th. After some useful sorties, the lake was bombed by a flight of Heinkel He 111s and N5628 sank through the shattered ice, lying on the bottom, until it was salvaged thirty years later.

Gloster Meteor IV EE549

see photo 54

Built 1946, single seat fighter, powered by two Rolls-Royce Derwent V turbojets specially rated to 4,200 pounds static thrust each.

EE549 was built at Hucclecote as a standard Mk. IV but was quickly converted to 'Star Meteor' status to join the Royal Air Force High Speed Flight (part of Fighter Command) at Tangmere. The modifications varied from higher rated engines, to a stronger canopy. It was delivered fully converted to Tangmere on August 1st 1946. On September 7th, flown by Group Captain E. M. Donaldson D.S.O., A.F.C., it took the absolute air speed record to 615·78 miles per hour over Rustington in Essex. Flushed with success, EE549 returned to

Glosters on October 3rd for preparation to be displayed at the Paris Salon in November. After the show it was flown Le Bourget-Croydon in the record time of twenty minutes eleven seconds - an average speed of 618·4 miles per hour, over the distance of 208 miles, on January 16th 1947. On that date it returned to Glosters for more modifications, which included the fitting of short span wings, these being completed on May 22nd 1947. Seven days later it joined the Fighter Command Communications Squadron at Bovingdon, joining the Central Fighter Establishment at West Raynham on March 29th 1948. It received some sort of damage on July 2nd whilst serving with the CFE but this was not serious. On June 10th 1952 it was retired and taken to the Station Museum at Cranwell, where it was allocated the maintenance airframe number 7008M. On June 4th 1958 it was moved to Fulbeck for storage, moving on to St. Athan in 1967. Here it was restored by No.19 Maintenance Unit to its original 'Star Meteor' status, and given the long-span wings again. It was briefly displayed at Horse Guards Parade for Battle of Britain Week in September 1968, before returning to St. Athan. It was delivered to Hendon in 1971.

Hawker Cygnet G-EBMB

see photo 52

Built 1924, constructor's number 2, two seat ultra-light, one Bristol Cherub III.

The Cygnet was designed by W. G. Carter and Sidney Camm for the 1924 Lympne Light Plane Trials. G-EBMB was the second Cygnet and was originally fitted with an Anzani twin cylinder engine of 1,100 c.c. It was flown at the trials as No.14. In 1926 it was fitted with a Bristol Cherub III and succeeded in winning the 1926 Competition. It was put into store at Brooklands in 1927, and emerged again in 1948, being made airworthy at Langley. It was donated to the Museum by Hawker Siddeley Aviation in 1972 and forms the centre piece of the Sidney Camm Memorial.

Hawker Hart 'J9941'

see photo 55

Built 1931, 'constructor's number' HH.1, light bomber, powered by one Rolls-Royce Kestrel IIIS piston engine of 560 horse power.

This aircraft was used by Hawkers as a development and demonstrator aircraft, the constructor's number being 'one off'. It was given the civilian registration G-ABMR and was allocated its first Certificate of Airworthiness on August 1st 1931, then fitted with a Rolls-Royce Kestrel 1B of 525 horse power. After the War it was refurbished in 1949 and given the engine noted in the heading and a blue and gold colour scheme. In this guise it won the 1952 Kings Cup race at a speed of 170 miles per hour. In 1959 it was refurbished and painted in No.57 Squadron colours as 'J9933' and based at Dunsfold with Hawker Siddeley's 'circus' of historic types. It was amended to J9941 in 1971. It was grounded in that year and presented to the Museum in 1972.

Hawker Hart Trainer IIA K4972

see photos 65 & 66

Built 1935, advanced trainer, powered by one Rolls-Royce Kestrel X piston engine.

K4972 was built by Armstrong Whitworth Aircraft and was first used by No.2 Flying Training School at Digby, later passing on to use at Brize Norton with an unknown unit. It was placed in store in 1938 and was to spend the early War years in a succession of Maintenance Units until 1943. In that year it was given the maintenance airframe number 1764M and was issued to No.1546 Squadron, Air Training Corps, for use as an instructional airframe. It was stored in the later 1940s and rediscovered dismantled in a barn near Carlisle by the Solway Aviation Group. It was in good condition considering its many years of storage. It was presented to the Royal Air Force Museum and its refurbishing was undertaken by No.19 MU at St. Athan, being completed by 1968. It was brought to Hendon in 1972.

Hawker Afghan Hind BAPC.82

Built circa 1937, two seat light bomber, powered by one Rolls-Royce Kestrel of 640 h.p.

This aircraft was a gift to the Royal Air Force Museum by the people of Afghanistan in 1968. It had been in service with the Afghanistan Air Force, and had not been broken up but stored in the open. It was flown home in an RAF transport and refurbished to its current splendour,complete with Afghan markings. Its past is not known, though it is possible that it was once K4672. It has been allocated the British Aircraft Preservation Council identification number 82. Further details on this aircraft appreciated.

Hawker Hurricane I P2617 'AF-T'

see photo 68

Built 1940, single seat fighter, powered by one Rolls-Royce Merlin piston engine of 1,280 horse power. (Original engine number 119697)

This Hurricane was built by Glosters and was initially delivered to No.20 Maintenance Unit at Aston Down on January 19th 1940, moving on to No.6 MU at Brize Norton on February 24th for preparation for service. The official history records that it was issued to No.615 Squadron *and* No.1 Repair and Salvage Unit on April 14th 1940. The following day it was issued to No.607 Squadron, and quickly suffered a Category B (beyond repair on site) flying accident, moving to No.4 MU at Ruislip on May 29th for repair. P2617 was to miss the Battle of Britain whilst under repair, being issued to No.1 (Canadian) Squadron on completion on November 21st 1940, going to Glosters on November 21st. It was issued to No.15 MU at Wroughton on March 29th 1941 and on to No.5 MU at Kemble from May 4th. P2617 then entered a phase of flying training flying, with No.9 Flying Training School from July 31st 1941, No.9 Service Flying Training School from December 6th, No.8 FTS from January 10th 1942, and back with No.9 FTS from March 9th. With No.9 FTS it suffered a Category Ac (repair beyond unit capacity) flying accident on May 9th, a team from Hawkers moving in on the 14th to effect a repair. It was back on charge with No.9 FTS on the 23rd and served peacefully until being retired to No.22 MU at Silloth on August 21st 1943. It was moved on to No.52 MU at Cardiff on August 6th 1944 and later to No.76 MU at Wroughton for 'museum storage'. It was refurbished by No.71 MU Bicester in the 1960s and used for several years as one of their travelling display airframes, until being allocated to the Royal Air Force Museum and moving to Hendon in 1972.

Hawker Typhoon IB MN235

Built 1943, single seat fighter, powered by one Napier Sabre piston engine.

MN235 had little operational life in the United Kingdom, being shipped to the United States in 1944 for evaluation by the United States Army Air Force as T-2/491. After testing and evaluation, it was handed over to the Smithsonian Institute, Washington and placed in store. As the only surviving Typhoon, the Royal Air Force Museum entered negotiations with the Institute and secured the aircraft for the Museum, in exchange for a Hurricane. It was returned to the United Kingdom and refurbished to display condition by No.27 Maintenance Unit at Shawbury. Moving to Henlow in 1970 and from there to Hendon.

Hawker Tempest V NV778

Composite rebuild, based on 1944 original, single seat high speed target tug, powered by one Napier Sabre IIa piston engine of 2,180 horse power.

This aircraft on display at Hendon has its origins in a pile of Tempest remains at the Proof and Experimental Establishment at Shoeburyness in 1958. From these No.33 Squadron builtone aircraft, which was displayed at the unit's standard presentation at Middleton St. George on April 24th 1958. It was marked up as 'SN219' and coded '5R-F'. 'SN219' moved on to Leeming. From here, the aircraft was earmarked for the Royal Air Force Museum, being the only Mark V in existence. It was moved to No.27 Maintenance Unit at Shawbury for refurbishing, and here it was found that parts of NV778 had been used in the rebuild, the aircraft was accordingly renumbered. It appeared at the Abingdon Review of that year, and at Horse Guards Praradein September. After this it was moved to Cosford, and then to No.71 MU at Bicester in 1972 for preparation for display at Hendon later that year. NV778 was built as a Mark V Series 2 Tempest at Langley in 1944 and was used for most of 1945 by Napiers, probably based at Luton. It was stored at No.5 MU, Kemble, until returning to Hawkers in March 1950 for conversion to a Tempest TT. Mk.5 target tug. It was stored at No.6 MU Brize Norton until September 1951 when it was moved to No.20 MU at Aston Down. It was issued to No.233 Operational Conversion Unit at Pembrey on October 27th 1952, serving here until July 1955 when it was sold to the Ministry of Supply who moved it to the wastes of Shoeburyness for testing.

Hawker Sea Fury FB. Mk.11 VX653

Built 1948, single seat naval fighter bomber, powered by one Bristol Centaurus 18 radial piston engine of 2,480 horse power.

This aircraft is part of the Sidney Camm Memorial and was transferred from the Fleet Air Arm in 1971. It was originally delivered to Royal Naval Air Station Anthorn in 1948. Its service history is unknown as this book closes for press. It was in store at Lossiemouth for most of the 1960s and it was from that station that VX653 was issued, refurbished, for display at Hendon.

Hawker Hunter F. Mk.5 WP185

Built 1955, single seat fighter, powered by one Armstrong Siddeley Sapphire turbojet.

WP185 was built by Armstrong Whitworth Aircraft at Baginton, and first flew in June 1955. It saw service with No.1 Squadron, as 'P', followed by No.34 Sqn, as 'E', and finally with No.208 Squadron. The Mk.5 Hunter was built as an insurance for the Avon powered versions, and the success of that type led to the Mk.5s being retired early, WP185 being allocated to Henlow on August 21st 1958 and stored for the Museum. It was allocated the maintenance airframe number 7583M at this time.

Hawker P.1127 XP831

Built 1960, prototype vertical take off fighter, powered by one Bristol Siddeley Pegasus 3 vectored thrust turbofan of 13,500 pounds static thrust.

XP831 was the first prototype P.1127, and was the first step to the Harrier 'jump-jet'. It made its first tethered hover at Dunsfold on October 21st 1960, graduating to its first free hover on November 19th. For these early trialsit was fitted with a Pegasus 1 of 10,400 pounds static thrust. For its first free flight, on March 13th 1961, it was fitted with a Pegasus 2 of 12,000 pounds static thrust. It was used for development work with Bristol Siddeley and with Hawkers for hover control and stabilisation trials. In 1962 it gained the powerplant named in the heading and was used for trials on the aircraft carrier H.M.S. 'Ark Royal' on February 8th 1963 - foreshadowing the Sea Harrier by fourteen years. It was displayed at the 1963 Salon at Le Bourget, but sadly suffered a flying accident, in front of its French rival, the Mirage Balzac. It was repaired and issued to the Aerodynamics Flight of the Royal Aircraft Establishment at Thurliegh, with whom it served until moving to the Royal Air Force Museum at Hendon by road on November 13th 1972.

Miles M.14A Magister T9707

Built 1940, constructor's number 1995, primary trainer, powered by one De Havilland Gipsy Major in-line piston engine of 130 horse power.

This aircraft served the Royal Air Force as T9707 until being demobbed and sold off to the civil market as G-AKKR. Magisters were known as Hawk Trainer IIIs in civil use. It flew with several private owners until being officially being withdrawn fron use in April 1965. It was acquired by the Royal Air Force Museum some time after this, and appeared in its military guise again at the Abingdon Review in 1968. It was installed in the Museum in 1972.

Royal Aircraft Factory SE.5A F938

Built 1918, constructor's number 687/2404, single seat fighter, powered by a Wolseley W.4A Viper piston engine of 200 horse power.

This aircraft was built as F937 by Wolseley Motors at Adderley Park, Birmingham in 1918. It was not used at all by the Royal Air Force. It was registered as G-EBIC to Major J. C. Savage on April 17th 1924 and converted by his company to one of the famous skywriting aircraft. It was withdrawn from use in September 1930 and was stored at Brooklands as part of the Nash Collection. It was restored at Colerne in 1950 as 'B4563' and was transferred to the Royal Aeronautical Society in December 1953, and ultimately to the Royal Air Force Museum on permanent loan. It was refurbished again by British European Airways at Heathrow in 1959 as an exercise for its apprentices. In 1968 the SE.5A was moved to Henlow and painted up as F938 and installed in the Museum in 1972. N.B. Identification clarification requested.

Short S.25 Sunderland MR. Mk.V ML824

see photo 53

Built 1944, long range maritime patrol flying boat, powered by four Pratt and Whitney R-1830-90D Twin Wasp radial piston engines of 1,200 horse power each.

ML824 was laid down on the line at Sydenham as a Mark III, but was completed as a V in July 1944. It was delivered to No.57 Maintenance Unit at Wig Bay on November 6th 1944. It was delivered to Calshot on February 9th 1945 and was taken on charge with No.201 Squadron two days later at Castle Archdale, here it was given the code letters 'NS-Z' that it wears today. Here it was used for anti U-boat patrols. On April 17th 1945 it was transferred to No.330 (Norwegian) Squadron at Sullom Voe in the Shetlands, and it continued to fly anti U-boat patrols. It was struck off charge on September 19th 1945, but was brought back on charge again with the Berlin Blockade on May 20th 1948 at Wig Bay. It was readied for possible use in flying supplies into Berlin from Finkenwerder, but was not used. In August 1949, after the Airlift had been completed it was given -90D Twin Wasps instead of the previously fitted -65s. ML824 was flown back to Sydenham on August 25th 1950 and was modified & refurbished for the French Naval Air Service - the Aeronavale. It was air tested on October 24th, flown to Pembroke Dock on the 26th, and on to Brest on the 28th. With the Aeronavale it was to serve from Dakar, with Flotilles F7, 12S, 27F and 50S. It was in Africa for all its life with the French, apart from a period on overhaul with Shorts at Sydenham from September 1956 until October 3rd 1956. In 1960 the Sunderlands were nearing the end of their useful life and Mr. Peter Thomas, later Director of the Skyfame Aircraft Museum, established the Short Sunderland Trust to save one of these aircraft and display it to the public. The Aeronavale donated ML824 to the Trust and it made its last flight from Brest to Pembroke Dock on March 23rd 1961. It had reached some 2,900 hours in the air at this time. It was open to the public during the summer months for ten years until handed over to the RAF Museum, as they did not have an example of the type. It was dismantled at Pembroke Dock and transport by a combination of road and sea to Hendon on March 21st 1971 and installed in the Grahame-White Hangar. It was refurbished and rolled out in its original No.201 Squadron colours in late 1976. It is planned to build a hangar around this magnificent exhibit in the near future.

Sikorsky VS-316 Hoverfly I KK995 'E'

see photo 59

Built 1944 originally, exhibit being a composite airframe, two seat light helicopter.

The aircraft on display is a composite of two aircraft used by the College of Aeronautics at Cranfield. Their histories follow. KK995 was delivered to the United Kingdom in December 1944 to Martin Hearn and Company at Hooton Park , who were responsible for the type in the U.K. It was then delivered to No.43 Operational Training Unit at Andover. It was issued to No.48 Maintenance Unit at Hawarden in February 1946 and back to Hearn's in October. From Hooton Park it moved to No.29 MU at High Ercall and then to the General Aircraft Company in November 1948. GAL used KK995 to repair KK987 and KL110, the remaining airframe going to the College of Aeronautics at Cranfield and it was used there for various experiments until being broken up in the later 1950s. KL110 was imported to Hearn's in February 1945 and was issued to No.48 MU at Hawarden. It was with the Andover-based No.43 OTU from October 1945 until February 1946, when it moved on to join the Transport Command Development Unit until going to No.29 MU in December 1947. It was with General Aircraft in 1948 and swallowed parts of KK995 in the process. It was issued to the King's Flight in July 1948 and was retired to the Royal Naval Air Station at Gosport in January 1949. It was used by the College of Aeronautics as a rotor rig in 1951/52. The combined parts (mostly KK995) were acquired by the Royal Air Force Museum in the early 1960s.

Sopwith 1.F1 Camel F6314

see photo 63

Built circa 1917, single seat fighter, powered by one Bently radial piston engine.

The service history of this aircraft is unknown. It was built for the Royal Flying Corps by Boulton and Paul Aircraft Limited and was acquired post-War by R. G. J. Nash and stored at Brooklands with the rest of his collection of historic aircraft. After the Second World War it was acquired by the Royal Aeronautical Society and later passed on to the Royal Air Force Museum on loan. It is displayed in the colours of No.65 Squadron.

Sopwith Triplane N5912

Built 1917, single seat fighter, powered by one Clerget rotary piston engine.

N5912 was one of only three built by the Oakley and Company works at Ilford, and was delivered to the Royal Flying Corps on October 19th 1917. It was first in service with the School of Aerial Fighting at Marske during 1918. In 1919 the School moved to Redcar. In 1936 the remains of this aircraft were discovered on the dump at Cardington and it was restored by Flight Lieutenant Buckle and Flight Sargent Scott for the 1936 Hendon Pageant. Its history from then on is unknown until it appeared in storage at No.5 Maintenance Unit at Kemble in 1945. In 1950 it was taken to No.39 MU at Colerne for refurbishing, and it was displayed at Hendon in 1951 before being stored at Fulbeck with historic aircraft. It was renovated by Hawker Siddeley Aviation at Dunsfold in 1961 and took part in the 1962 Royal Tournament at Earls Court. After this it was taken to Biggin Hill for storage, and was displayed for a time in the Fleet Air Arm Museum at Yeovilton until going to Hendon in 1972.

Supermarine Stranraer 920 'QN'

Built 1940, constructor's number CV-209, maritime patrol flying boat, powered by two Wright Cyclone GR-1820-G202A radial piston engines.

920 was built as such for the Royal Canadian Air Force by Canadian Vickers, and was fitted with Bristol Pegasus XXII engines. It was delivered to RCAF Eastern Air Command on November 28th 1940 and served with No.5 (BR) Squadron. It was transferred to Western Air Command at

Patricia Bay on October 1st 1941. It was sent to Boeings at Vancouver on January 23rd 1942 for overhaul, being re-delivered to WAC on June 13th. 920 then went on to serve with Nos. 7 (BR) and 9 (BR) Squadrons. It was delivered to Canadian Pacific Airlines at Vancouver on September 16th 1943 following a Category B accident. It was repaired and re-delivered to WAC on February 1st 1944 for a short period before going to No.3 Repair Depot on April 12th for storage. It was sold to the Labrador Mining and Exploration Company Limited on May 10th 1944 and given its Certificate of Airworthiness, as CF-BXO, on May 24th. It was sold to E. Lando of Vancouver on April 23rd 1947 and to Queen Charlotte Airlines Limited on May 28th. In March 1950 it was re-engined with the engines described in the above title. It was damaged in an accident at Sullivan Bay on February 23rd 1952 and stored, being sold to W. K. Slye on December 15th 1958. It was registered to Stranraer Aerial Enterprises Limited of Vancouver on June 28th 1962 and was given a new C. of A. In 1970 it was bought by the Royal Air Force Museum and freighted back to the United Kingdom in the hold of a Belfast C. Mk.1 on August 12/19th 1970. It was moved into the Museum at Hendon in 1972 and has to be displayed without its beaching gear wheels, as fitting these would put it through the roof! It is displayed in its Royal Canadian Air Force colours.

Supermarine Spitfire Ia K9942 'SD-V' see photo 67

Built 1939, single seat fighter, powered by one Rolls-Royce Merlin II piston engine.

This aircraft has an impressive War record behind it. It was first delivered to No.72 Squadron, then at Church Fenton, on April 24th 1939, joining 'A' Flight of that unit. The Squadron moved about considerably during the next year. In October it moved to Leconfield, followed by Drem, a return to Church Fenton, to Acklington on March 2nd 1940 and to Gravesend on June 1st 1940. On June 5th 1940, after a patrol over Dungeness at 20,000 feet, K9942 landed back at Gravesend with its undercarriage up, suffering Category B (damage beyond repair on site) damage. It was on charge with No.4 Maintenance Unit at Ruislip from June 21st and was repaired by General Aircraft Limited there, it was sent to No.6 MU at Brize Norton before joining No.57 Operational Training Unit at Hawarden on August 17th. K9942 served with No.57 OTU without incident until February 6th 1942 when it was again declared Category B, repairs were started on the 13th, and it was ready for collection again on September 10th, going to No.33 MU at Lyneham seventeen days later. On April 10th 1943 it was on charge with No.53 OTU, based at Kirton-in-Lindsey. Here again, K9942 was the victim of a flying accident, again Category B, on October 22nd 1943. It was repaired and again issued to No.33 MU, on March 3rd 1944, moving to No.82 MU at Lichfield on May 15th. On August 28th it was earmarked for museum use and was put into the care of No.52 MU at Cardiff. It moved to No.58 MU at Newark in September 1951 and was later to appear in static displays at Horse Guards Parade during Battle of Britain Week. It later was on charge with No.71 MU at Bicester who used it as one of their travelling demonstration airframes, painted in the superb 1939 colours of No.72 Squadron, codes 'SD-V', that it wears today. K9942 was moved to Hendon on November 9th 1971.

Supermarine Spitfire F. Mk.24 PK724

Built 1946, single seat fighter, powered by one Rolls-Royce Griffon piston engine.

PK724 has no operational service history at all, and has about seven hours total flying time on the airframe! It was delivered to No.33 Maintenance Unit at Lyneham on October 30th 1946 and remained there until 1949. On April 5th 1949 it was entrusted to Westland Aircraft for modifications, moving nine days later to a location thought to be in Scotland for further modifications, this work was completed and the aircraft was ready for collection on February 2nd 1950. It was issued to No.9 MU at Cosford a week later. It was declared a non-effective airframe on December 14th 1954 and allocated the maintenance airframe number 7288M on October 31st 1955. From November 4th 1955 it was moved to Norton for display work, followed by a period on the gate at Gaydon from December 1st 1961. It was moved to Henlow on March 6th 1967 for possible use in the film 'The Battle of Britain', returning to Gaydon on August 13th 1968. It was moved to Finningley on February 20th 1970 for preparation for display in the Royal Air Force Museum, being moved to Hendon on April 2nd 1971.

Vickers FB.5 Gunbus replica '2345'

Built 1966, constructor's number VAFA.01, two seat fighter replica, powered by one Gnome Monosoupape rotary piston engine driving a pusher propeller.

This aircraft is an exact flying replica, built by the Vintage Aircraft and Flying Association, at Weybridge. It was given the civil registration G-ATVP, and the Royal Flying Corps serial 2345 appeared on the tail. It was fitted with an original engine, and flew for the first time from Weybridge on June 14th 1966. It was flown for a short while, and then presented to the Royal Air Force Museum and stored at Henlow before moving into Hendon.

Vickers FB.27A Vimy replica 'F8614'

Built 1969, constructor's number VAFA.02, medium bomber replica, powered by two Rolls-Royce Eagle VIII piston engines of 360 horse power each.

This was the second aircraft to be built by the Vintage Aircraft and Flying Association, the Vimy being built to celebrate the 50th Anniversary of the Alcock and Brown Transatlantic Flight of 1919 - see under Science Museum for more details. It was built at Weybridge and, in usual VAFA tradition, followed original lines as closely as possible, and used original engines. It flew for the first time, unmarked although allocated G-AWAU, on June 3rd 1969. It was soon painted as H651 and went on display at Ringway Airport in July. Sadly, the intense summer of 1969 caused the fabric of the lower wings to ignite on July 16th and the aircraft was badly damaged in the following fire. It was rebuilt to static condition and donated to the Royal Air Force Museum, being painted up as 'F8614', in Nivo green finish and carrying the name 'Triple First'.

Vickers Wellington T. Mk.10 MF628 see photo 60

Built 1944, medium bomber, powered by two Bristol Hercules XVI radial piston engines.

MF628 was built at Squires Gate, and was delivered to No.18 Maintenance Unit at Dumfries on May 11th 1944. It was stored for what remained of the War. In March 1948 it was on charge with Boulton and Paul for conversion into a T. Mk.10 crew-trainer. This work completed, it was allocated to No.1 Air Navigation School, at Hullavington, on April 13th 1949. It was given the code letters 'FFK-B'. It was involved in a serious accident, Category 4 (beyond repair on site) in December 1951 and was taken to Brooklands for repair by Vickers. This work was finished in October 1952, but MF628 had been overtaken by the introduction of the Valetta T. Mk.3 to No.1 ANS and, instead of going back into service, it was flown to No.19 MU at St. Athan. It was declared a non-effective airframe on January 8th 1953, but was kept airworthy as it appeared in the flying display at the 'Fifty Years of Flying' display at Hatfield on June 14th 1953. It flew again at the St. Athan Battle of Britain Display that September. It was flown to Hemswell in March 1954 to take part in the film 'The Dam Busters'. It flew 'home' to St. Athan and was again declared non-effective in October 1954. On January 24th 1955 it was sold to Vickers who presented it the following year to the Royal Aeronautical Society, and it appeared as a static exhibit at the Garden Party of the R.Ae.S. at Wisley on July 15th 1956. It was moved to Hendon in 1957 for storage, and then to Heathrow in 1958/9 for storage with British European Airways Maintenance Base, along with the Nash Collection. MF628 was presented to the Royal Air Force in 1964, by now the only surviving 'Wimpey'. It went to St. Athan again in 1968 for refurbishing prior to being displayed at the Abingdon Review of June 1968. It moved to Henlow in July. It was moved into Hendon on October 26th 1971, it is hoped to eventually restore it to its turreted B. Mk.X days.

Westland Lysander III R9125 'LX-L' see photo 58

Built 1940, two seat army co-operation/general duties aircraft, powered by one Bristol Mercury XVA radial piston engine.

The service history of this 'Lizzie' is not well documented as to its whereabouts for a large stretch of the War. It was delivered to No.225 (Army Co-operation) Squadron at Tilshead on September 29th 1940. It was used for coastal patrol and for photo reconnaissance work. On April 29th 1941 six Lysanders arrived at the unit to replace a similar number of aircraft that were found to have cracked engine mounting struts. It is assumed that R9125 was one of these aircraft and that it was removed to a contractors works for repair. It did not appear again until October 15th 1944 when it was allocated to No. 161 Squadron at Tempsford. It is not recorded as having done any operational flying with this unit. After this date the aircraft is thought to have been stored and soon allocated to the Air Historical Branch as a significant airframe. It appears in the Museum wearing the original colours and codes of its days with No.225 Squadron.

<u>Notes</u> : The following incomplete airframes are also on display inside the Museum :-

<u>Armstrong Whitworth R.33 Airship G-FAAG</u> section of control gondola only, scrapped 1927.
<u>Bristol 96 F.2b Fighter</u> fuselage frame only, from Weston-on-the-Green.
<u>Bristol 156 Beaufighter</u> cockpit section only, from Cranfield.
<u>Morane 'BB' A301</u> fuselage frame only.

Royal Aircraft Factory BE.2 fuselage frame only.

The following aircraft are housed in the following areas of the Museum :-

Main Hall : Avro 504K E449, Avro Lancaster I R5868, Bleriot Type XI No.164, Boulton Paul Defiant I N1671, Bristol Beaufighter TF. Mk.10 RD253, Bristol Belvedere HC. Mk.1 XG474, Caudron G.III '3066', Cayley Glider replica BAPC.89 (suspended from the ceiling), English Electric Canberra PR. Mk.3 WE139, English Electric P.1B XA847, De Havilland Tiger Moth II T6296, De Havilland Mosquito T. Mk.III TW117, Gloster Gladiator II K8042, Gloster Gladiator II N5628, Gloster Meteor IV EE549, Miles Magister T9707, Royal Aircraft Factory SE.5A F938, Sikorsky Hoverfly I KK995, Sopwith Camel F6314, Sopwith Triplane N5912, Supermarine Stranraer 920, Supermarine Spitfire IIa K9942, Supermarine Spitfire F. Mk.24 PK724, Vickers Gunbus replica '2345', Vickers Vimy replica 'F8614', Westland Lysander III R9125.
Sidney Camm Memorial Hawker Cygnet G-EBMB, Hawker Hart J9941, Hawker Hart Trainer K4972, Hawker Afghan Hind BAPC.82, Hawker Hurricane I P2617, Hawker Typhoon IB MN235, Hawker Tempest V NV778, Hawker Sea Fury FB. Mk.11 VX653, Hawker Hunter F. Mk.5 WP185, Hawker P.1127 XP831.
Gallery Displays All the incomplete airframes noted above.
Dermot Boyle Wing De Havilland Canada DHC-1 Chipmunk T. Mk.10 WP912.
Outside Main Building Blackburn Beverley C. Mk.1 XH124, Short Sunderland MR. Mk.5 ML824.

The Royal Air Force Museum has a large reserve collection, from which new exhibitions at Hendon, and elsewhere, will be made. The following is a brief listing of aircraft currently on charge with the reserve collection. The reader is reminded that the collection at Cosford is also part of the Royal Air Force Museum reserve collection, and that reference should be made to that Museum under the heading 'Cosford Aerospace Museum' for the complete picture. Two aircraft with No.2 SoTT at Cosford for the Museum are listed.

Airspeed AS.40 Oxford I MP425 ex G-AITB from Scone	kept at Cardington
Airspeed AS.40 Oxford I G-AITF ex ED290, from Scone, used for spares	kept at Henlow
Auster C4 T. Mk.7 'Antarctic' WE600 Antarctic Expedition a/c, 7602M	kept at Finningley
Avro 504K 'H2311' ex Nash Collection and R.Ae.S. ex G-ABAA	kept at Henlow
Avro 652A Anson C. Mk.19/2 TX214 7817M *see photo 73*	kept at Henlow
Avro 716 Shackleton MR.3/3 WR974 8117M 'K' in use with No.2 SoTT	kept at Cosford
Avro 716 Shackleton MR.3/3 WR977 possibly for disposal	kept at Finningley
Avro 707A WZ736 7868M, not fitted with undercarriage *see photo 62*	kept at Finningley
Beagle B.206 Basset CC. Mk.1 XS770 active with A&AEE	kept at Boscombe Down
Blackburn B.101 Beverley C. Mk.1 XL149 7988M 'X', for disposal?	kept at Finningley
Bleriot Type XI No.16	kept at Henlow
Bleriot Type XXVII No.433	kept at Cardington
Boulton Paul Sea Balliol T. Mk.21 WL732 ex A&AEE hack	kept at Henlow
Bristol 96 F.2b Fighter fuselage frame only, ex Weston-on-the-Green	kept at Henlow
Bristol 96 F.2b Fighter fuselage frame only, ex Weston-on-the-Green	kept at Henlow
Bristol Bolingbroke Mk. IVT 10001 ex R.C.A.F., under restoration	kept at Boscombe Down
Bristol 171 Sycamore HR. Mk.12 WV783 7841M	kept at Henlow
Bristol 171 Sycamore HR. Mk.14 XL824 8021M	kept at Henlow
Bristol 173 Series 1 XF785 7648M, c/no. 12871, ex G-ALBN	kept at Henlow
Bristol 192 Belvedere HC. Mk.1 XG454 ex 66 Squadron	kept at Henlow
CASA 2.111 forward fuselage, used in 'Battle of Britain' film	kept at Henlow
De Havilland DH.89 Dragon Rapide G-AHED under restoration to RL962.	kept at Cardington
De Havilland DH.98 Mosquito TT. Mk.35 TJ138 7607M 'VO' ex Colerne	kept at Finningley
De Havilland DH.106 Comet 1XB XM823 c/no. 06022, ex F-BGNZ, G-APAS	kept at Shawbury
De Havilland DH.106 Comet C. Mk.2 XK699 7971M, ex Lyneham	kept at Henlow
De Havilland DH.100 Vampire FB. Mk.5 VX461 7646M, ex Hendon	kept at Henlow
De Havilland DH.112 Venom NF. Mk.3 WX905 7458M, ex Hendon	kept at Henlow
De Havilland DH.115 Vampire T. Mk.11 XD506 'E' ex CATCS	kept at Finningley
De Havilland DH.115 Vampire T. Mk.11 XE920 8196M, 'D' ex CATCS	kept at Henlow
De Havilland Canada DHC-1 Chipmunk T. Mk.10 WP861 ex Cambridge UAS	kept at Kemble
EoN Eton TX.1 WP270 primary glider	kept at Henlow
English Electric Canberra B. Mk.2 WD935	kept at St. Athan
English Electric Canberra B. Mk.2 WJ573 'Operation Swifter' aircraft	kept at Henlow
English Electric P.1A WG763 7816M, on show on the parade ground	kept at Henlow
Fairchild Argus II G-AIZE c/no. 565, ex 43-14601, N9996F	kept at Henlow

63 *Camel F6314 sunning itself at Abingdon 1968*

64 *Hawk Major DG590 at Henlow May 1970* Stuart Howe

65 *Before...Hart K4972 at Carlisle just after discovery*

66 *After....refurbished and on show at Biggin Hill 1969*

67 *Spitfire K9942 at Wavertree in 1970* S. G. Jones

68 *Hurricane P2617 also at Wavertree, Liverpool* S. G. Jones

69 *Fokker Dr.I 8417 at Cranfield 1966* P. H. Butler

70 *Bf 109 4101 at Horse Guard's Parade 1955* P. H. Butler

71 *He 111 701152 on parade, 12/9/55* P. H. Butler

72 *Ju 88 360043 also at Horse Guard's Parade* P. H. Butler

73 *Anson C.19 TX214 at Henlow May 1970* Stuart Howe

74 *Seagull VH-ALB in bad way at Bankstown* R. A. Scholefield

Fairey Battle I L5343 remains from Iceland, under restoration kept at Leeming
Fairey Swordfish II HS503 from Canada kept at Henlow
Farman F.40 F-HMFI under restoration kept at Cardington
Fiat CR.42 MM5701 ex BT474 kept at St. Athan
Fieseler Fi 156C-1 Storch 475081 ex VP546, 7362M, under restoration kept at St. Athan
Focke-Achgelis Fa 330A-1 identity unknown kept at Henlow
Focke-Wulf Fw 190A-8/U1 584219 two seat version kept at St. Athan
Fokker D.VII 2319/18 *see photo 69* kept at Henlow
General Aircraft Hotspur II HH379 rear fuselage, marked '27' kept at Henlow
General Aircraft Hamilcar NX836 forward fuselage section only kept at Henlow
Gloster Meteor F. Mk.8 WH301 7930M, 'T' ex Manby kept at Henlow
Gloster Meteor F. Mk.8 'WH456' really WL169, 7750M kept at Finningley
Gloster Meteor T. Mk.7/8 Hybrid WA634 ex 'MBA634' from Chalgrove kept at St. Athan
Gloster Meteor NF. Mk.14 WS843 7937M ex Henlow kept at St. Athan
Gloster Javelin FAW. Mk.4 XA549 7717M ex Swanton Morley kept at Finningley
Hafner Rotachute III P-5 single seat, strap-on, rotor-parachute kept at Henlow
Handley Page HP.61 Halifax II W1048 ex Norway, under restoration kept at Henlow
Handley Page HP.67 Hastings T. Mk.5 TG511 operational with No.230 OCU kept at Scampton
Handley Page HP.80 Victor K. Mk.1A XH593 in use with No.2 SoTT kept at Cosford
Hawker Tempest V EJ693 'SA-J' fuselage only, from the Netherlands kept at Henlow
Hawker Hunter F. Mk.3 (P.1067) WB188 7154M, ex Colerne kept at St. Athan
Hawker Hunter F. Mk.1 WT619 7525M, sectioned display, ex St. Athan kept at Henlow
Hawker Siddeley Kestrel FGA. Mk.1 XS695 A2619, with wings of XS696 kept at Manadon
Heinkel He 111H-23 701152 ex Henlow *see photo 71* kept at St. Athan
Heinkel He 162 120227 ex VH513, ex Colerne kept at St. Athan
Hunting P.84 Jet Provost T. Mk.1 XD674 7570M, prototype kept at Finningley
Junkers Ju 87D-3/Trop 494085 'W8-AE' kept at St. Athan
Junkers Ju 88R-1 360043 'DS-EV' ex PJ876 and Henlow *see photo 72* kept at St. Athan
Messerschmitt Bf 109E-3 4101 ex DG200 '12' *see photo 72* kept at St. Athan
Messerschmitt Bf 109G-2 10639 under restoration kept at Lyneham
Messerschmitt Bf 110G4/R3 730301 kept at St. Athan
Messerschmitt Me 163A-2 191904 '25', ex Colerne kept at Finningley
Messerschmitt Me 262A-1 112372 'X' kept at Finningley
Miles M.2H Hawk Major DG590 ex G-ADMW, ex Ternhill ***see photo 67*** kept at Henlow
Miles M.38 Messenger 4A G-ALAH ex NAPS, for restoration as RH377 kept at Henlow
Mitsubishi Ki 46-III Dinah BAPC.84 ex ETPS '29' kept at St. Athan
Percival Proctor III Z7197 ex G-AKZN, ex Shawbury kept at Finningley
Percival P.56 Provost T. Mk.1 WV499 'P-3', 7698M kept at St. Athan
Percival P.56 Provost T. Mk.1 XF545 'P-Z', 7957M kept at Finningley
Saunders-Roe Skeeter AOP. Mk.12 XN341 8022M kept at St. Athan
Saunders-Roe SR.53 XD145 ex Westcott *see photo 61* kept at Henlow
Scottish Aviation Pioneer CC. Mk.1 XL703 8034M kept at Henlow
Scottish Aviation Twin Pioneer CC. Mk.1 XL993 kept at Henlow
Slingsby Cadet TX. Mk.1 identity unknown. kept at Henlow
Sopwith Dolphin D5329 remains under reconstruction kept at Cardington
Supermarine Seagull V A2-4 ex VH-ALB and Royal Australian Air Force kept at Wyton
Supermarine Southampton I N9899 hull only ex house-boat, ex Henlow kept at Cardington
Supermarine Spitfire I X4590 'PR-F' ex Henlow kept at Finningley
Supermarine Spitfire Vb 'BL614' really AB871, 'ZD-F', ex Colerne kept at St. Athan
Supermarine Spitfire IX MK356 5690M kept at St. Athan
Supermarine Spitfire F. Mk.21 LA255 6490M 'JX-U' kept at Wittering
Supermarine Spitfire F. Mk.22 PK624 'RAU-T' displayed on the gate kept at Abingdon
Supermarine Swift FR. Mk.5 WK281 'S', 7712M kept at Finningley
Vickers Valetta C. Mk.2 VX573 'Loreley' kept at Henlow
Vickers Varsity T. Mk.1 WF414 'CC' ex No.5 FTS kept at Finningley
Vickers Valiant BK. Mk.1 XD818 7894, displayed in the camp kept at Marham
Westland WS-51 Dragonfly HR. Mk.1 VX595 '29', ex Fleetlands kept at Henlow
Yokosuka MXY-7 Ohka II BAPC.98 ex Cottesmore kept at Henlow

The aircraft of the Battle of Britain Memorial Flight, based at Coningsby, are all officially part of the Royal Air Force Museum Collection, they are as follows :-

Avro 683 Lancaster I PA474 coded 'KM-B' in the colours of No.44 Squadron.
Hawker Hurricane IIc LF363 coded 'LE-D' in the colours of No.242 Squadron.

Hawker Hurricane IIc PZ865 coded 'DT-A' in the colours of No.257 Squadron, ex G-AMAU.
Supermarine Spitfire IIa P7350 coded 'UO-T', ex G-AWIJ.
Supermarine Spitfire Vb AB910 coded 'QJ-J' in the colours of No.92 Squadron, ex G-AISU.
Supermarine Spitfire PR.XIX PM631 coded 'AD-C' in the colours of No.11 Squadron.
Supermarine Spitfire PR.XIX PS853 in PRU blue.

ROYAL SCOTTISH MUSEUM

The Department of Technology of the Royal Scottish Museum started collecting aeronautical material in 1909 with Pilcher's Hawk glider and models of Wright Model A and a Bleriot Type XI. Then in the 1920s a good collection of aero engines was gradually built up including four rotaries, a 28 horse power Anzani and an engine from a Wright Model B - presented by Orville Wright. In the years following World War Two the engine collection was again extended.

At the present time the display consists of Pilcher's Hawk, a selection of models and several engines. The engines on display include the Anzani and the Wright mentioned above and two sectioned and motorised units - a 110 horse power Le Rhone and a Bristol Hercules. It is proposed to keep a representative collection of material on display in the Chambers Street building despite the opening of a Museum of Flight (which see) at East Fortune Airfield because the latter cannot be opened to the public all year.

Admission is free of charge, and the Museum is open to the public all year (except Christmas and New Year) from 1000 to 1700 hours.

Address : The Royal Scottish Museum, Chambers Street, Edinburgh EH1 1JF.

Telephone : 031 225 7534.

Pilcher Hawk BAPC.49

Built 1896, pioneering hang-glider.

This machine was the last of four hang-gliders designed and built by Percy S. Pilcher. Pilcher was killed whilst flying this example on September 30th 1896 due to structural failure.

ROYAL TANK MUSEUM

The famous Royal Tank Museum at Bovington sets out to show the public in vivid form the development of the armoured fighting vehicle from its first real usage on the Somme in 1916 up to the modern-day sophistication of battle-tanks. Examples of most types of tank can be found, set in a spacious open-air site. These range from the Type IV from 1916 - the famous lozenged shaped tank to which the code name 'Tank' was given and has stuck for vehicles of that type ever since. The Second World War is well represented with the Matilda, Churchill, Sherman, Grant, all leading on to the Centaur and the Centurian

Admission times and fees are not available as this book goes to press, further details from the address below.

The Museum has had a Skeeter air observation post helicopter on display since 1970 and it has recently been refurbished for a further session on display.

Address : The Royal Armoured Corps Tank Museum and Royal Tank Regiment Museum, Bovington Military Camp, Bovington, Wareham, Dorset.

Saunders-Roe Skeeter AOP. Mk.12 XM564

Built 1959, constructor's number S2/5119, aerial observation post helicopter, powered by one De Havilland Gipsy Major 215 in-line piston engine of 215 horse power.

XM564 was ready for collection by No.652 Squadron Army Air Corps at Eastleigh on November 30th 1959. It moved to No.12 Flight on May 7th 1960. In 1962 it was transferred to the Royal Air Force for service with the Central Flying School at Ternhill, coded 'X', this happening until 1964. It was returned to the AAC and issued to No.15 Maintenance Unit at Wroughton on December 13th 1964. It was back in service again from August 23rd 1965 with No.5 Field Regiment Royal Artillery. It was reported back with No.652 Squadron from June 1964 until February 1966. It was struck off charge on January 1st 1969 and moved to Bovington for display shortly afterwards. It was refurbished by the Junior Leaders Regiment at Bovington, during 1974/75 and is now back on display at the Museum.

ROYAL UMPIRE MUSEUM

The Royal Umpire Museum receives its name from the famous coach the 'Royal Liverpool Umpire'. The Museum has many attractions and facets, and is a place with something of interest to everyone, young and old. There is a replica village and village green. Apart from the 'Royal Liverpool Umpire' there are many other types of coaches and carriages, the best in the north of England. There is a replica of the old Croston Hall, using many of the salvaged artifacts of the original. The Gradwells Cottage is also part of the Museum, it is said to be haunted by a lady ghost, and that a passage leads from the house to the Croston Rectory. There is a fascinating series of other attractions, including a full size Western Town for the children to play amongst.

The Museum has been kind enough to allow The Aeroplane Collection (previously the Northern Aircraft Preservation Society) to display two of its aircraft on the site. At present these comprise a Vampire trainer and a Royal Navy Meteor.

The Royal Umpire Museum is on the A581 Chorley to Southport road at Croston.

Admission fees on application to the Museum. It is open to the public from March until November, from 0900 hours to dusk.

Address : The Royal Umpire Museum, Croston, near Chorley, Lancashire.

Telephone : Croston 600257.

De Havilland DH.115 Vampire T. Mk.11 XD535 '54'

Built 1954, constructor's number 15280, advanced trainer, powered by one De Havilland Goblin 35 turbojet of 3,500 pounds static thrust.

XD535 was built at Hawarden and was handed over to the Royal Air Force at No.10 Maintenance Unit at Hullavington on May 28th 1954. Its first active assignment was in West Germany with No.93 Squadron at Celle from July 23rd. Here it was coded 'B'. On July 23rd 1957 it returned to the United Kingdom and took up residence with No.1 Air Navigation School at Thorney Island. It was to stay with the trainer role for the rest of its life. On April 2nd 1959 it became part of No.3 Squadron of No.5 Flying Training School, based at Oakington, staying here until March 21st 1962 when it joined No .4 FTS at Valley, coded '54'. It was retired to No.19 MU at St. Athan and sold to Hawker Siddeley Aviation on November 28th 1967. Hawkers intended to refurbish and resell the aircraft, and meanwhile placed it in store at Hawarden, its birthplace. Sales prospects for the Vampire were not as good as at first thought (it is thought that HSA bought in many Vampires for the planned Mystery Jet executive conversion) and XD535 was eventually shuffled over to Woodford in 1971/72 for still more time in mothballs. In 1973 it was offered for preservatíon to the British Aircraft Preservation Council and it was taken up by The Aeroplane Collection and moved from Woodford to Croston on March 16th 1973.

Gloster Meteor T. Mk.7 WL332

Built 1951, advanced trainer, powered by two Rolls-Royce Derwent 8 turbojets of 3,600 pounds static thrust each.

WL332 was built at Hucclecote to a Royal Navy contract for use as advanced trainer and standards checker. It was to fly with the Fleet Air Arm in this guise until about 1965 when it was withdrawn from use. Throughout this time it served in its black colour scheme. It was sold off to the Technical College at Moston and used for them as an instructional airframe. It was acquired by The Aeroplane Collection Limited in 1974, in a swop with a Vampire T. Mk.11. WL332 was delivered by road from Moston to the Royal Umpire Museum on June 29th 1974.

SCIENCE MUSEUM

The Science Museum houses the National Aeronautical Collection which affords the layman and historian alike a unique opportunity to savour the atmosphere of the pioneer days of aviation and to study the technical developments which have produced the advanced aircraft of today.

Over twenty aircraft are exhibited, plus several nose sections and balloon baskets and gondolas, all this is backed up by more than 140 models, and numerous displays illustrating the more technical side of aviation, such as aerodynamics, experimental aircraft and instrumentation. The aero-engine collection is the finest on display anywhere in the world. At the present time over 100 engines and 40 propellers cover the history of propulsion over the last 130 years.

The first acquisition of aeronautical apparatus by the Science Museum was in 1896 when

Sir Hiram Maxim presented a model of his flying machine, together with one of the steam engines and 18 feet diameter propellors which briefly lifted the machine off its guiding track before crashing. It is greatly to the credit of the Museum curators that they had the foresight to preserve this early material at a time when it was not at all clear that powered flight was possible. Even the eminent scientist Lord Kelvin declared "I have not the smallest molecule of faith in aerial navigation other than ballooning".

The first aircraft to come to the Museum was the 1912 Cody Biplane which arrived in 1913. But it was not until after World War One that the collection really began to develop, a major landmark being the presentation in 1919 of the Vickers Vimy aircraft shortly after Alcock and Brown had flown it across the Atlantic. The aeronautical collection at the Science Museum became known as the National Aeronautical Collection in the mid 1920s; the first use of this title in a Museum publication being in a catalogue of 1929.

It was the reputation of the NAC which led Orville Wright to lend the original 1903 Wright Flyer to the Science Museum in 1928 following his dispute with the Smithsonian Institution in Washington who refused to accept the priority of the Wright brothers' 1903 flight claims. After the Smithsonian had settled their differences with Orville Wright, the original aircraft was returned to America in 1948 and an exact replica was built for the Museum by the apprentices of the De Havilland Aircraft Company. The reputation of the Museum also attracted those most technically advanced aircraft of their day, the Supermarine S.6B and the Gloster E28/39, which were presented in 1932 and 1946 respectively.

In the years immediately after World War Two, the Collection was displayed in unsuitable galleries in the old Imperial Institute, just to the north of the Science Museum. Coincident with the expansion of Imperial College, it was decided to complete the Science Museum Centre Block and rehouse the Collection in a new gallery on the top floor. The basement and the ground floor of this building had been constructed for the Science Exhibition of the 1951 Festival of Britain.

It was decided that the Vickers Vimy, one of the most important aircraft in the collection, should be displayed in its entirety, and the lay-out of the gallery was accordingly designed around this one exhibit. The gallery itself was designed to be reminiscent of an aircraft hangar and to permit the aircraft to be suspended from the ceiling, giving the display a life and character not possible with the aircraft on the ground. It also allowed more objects to be displayed and although the exhibits are somewhat crowded the Museum believes that it is better for the public to see as much as possible rather than just a few orderly and more spaciously displayed objects. The use of a walkway through the gallery allows the public to move freely among the aircraft.

When the move to the new gallery took place the aircraft were taken apart, crated and lifted by a 100 foot jib crane through the long window in the north west corner of the Gallery. Surprisingly, none of the exhibits gave any real trouble in spite of the 'Flying Bedstead' having only 1½ inches to spare!

Although the present space restrictions preclude the display of large, modern aircraft the Science Museum is still pursuing its policy of acquiring technically significant items. In 1976 the Science Museum acquired the British-built Concorde prototype 002. This aircraft has been lent to the Fleet Air Arm Museum, Yeovilton, where it is now on public exhibition.

The Museum is open from 1000 until 1800 hours on week-days and from 1430 until 1800 hours on Sundays. The Museum does not open on Christmas Eve, Christmas Day, Boxing Day and New Year's Day. Admission is free.

Address : The Science Museum, South Kensington, London SW7 2DD.

Telephone : 01 589 6371.

Avro 504K D7560

Built 1918, primary trainer, powered by one Clerget rotary engine of 130 horse power.

This aircraft was built by the A. V. Roe Company in 1918, as part of a large order for the Royal Flying Corps, it was originally equipped with a Le Rhone engine of 110 horse power. Its service history (if any) is unknown; it was delivered to the Museum from No.3 Aircraft Salvage Unit, Waddon, in January 1920.

Cody Biplane 304

Built 1912, single seat general purpose biplane, powered by one Austro-Daimler piston engine of 120 horse power driving a pusher propeller.

This aircraft was flown by Cody at the Cambridge military manouevers of September 1912 and is similar to the one flown at the Military Aircraft Trials of the previous August. It was bought using Army funds, and carried their serial number 304. It was retired after a very short life and presented to the Science Museum in 1913. It is the only Cody aircraft left in existence.

Cierva C.30A Rota I AP507 'KX-H'

Built 1934, constructor's number 728, single seat calibration autogiro, powered by one Armstrong Siddeley Genet Major 1A radial piston engine of 140 horse power.

Built by Avro as a series 671 C.30A under licence, this aircraft was registered on November 16th 1934 to E. J. Jobling as G-ACWP at Hanworth. By October 1938 it was with a Mr. Meikle and was registered to the Autogiro Flying Club at Hanworth in March 1939. It was impressed for War service on June 1st 1940, becoming AP507 in 1941. It was used initially by the Royal Aircraft Establishment at Farnborough, but was transferred to the Calibration Flight at Duxford on August 28th 1940. It became part of No.74 Wing, and was with them when it was renamed No.1448 Flight (based at Duxford) in February 1942 and later No.529 Squadron from June 15th 1943. At this time it acquired the codes 'KX-H'. It was loaned to the Admiralty for experiments with anti-submarine gear from October 6th to 11th 1943 at Sydenham. It was retired on May 18th 1945 and moved to No.5 Maintenance Unit at Kemble to await disposal. It was sold to the Science Museum and was moved to No.76 MU at Wroughton for preparation. It was stored at Sydenham until the Museum could accomodate it, and was refurbished to display standards by the apprentices at No.1 School of Technical Training at Halton in 1961. It was put on display in the new gallery at the Museum on July 10th 1963.

De Havilland DH.60G Gipsy Moth G-AAAH

Built 1928, constructor's number 804, two seat sporting plane, powered by one De Havilland Gipsy 1 in-line piston engine of 100 horse power, and fitted with extra tankage.

G-AAAH was built at Stag Lane as the fourth production Gipsy Moth and was built as a special long range version. It was used by W. L. Hope (of Air Taxis Limited) for a flight to Kisumu and back in September 1928. G-AAAH was to come to fame as the bottle green and silver mount called 'Jason' of Amy Johnson. Miss Johnson took off from Croydon on May 5th 1930 to become the first woman to fly from the United Kingdom to Australia (10,000 + miles) arriving at Darwin on May 24th of that year. This was a fantastic achievement for any pilot, let alone a woman, and Amy became a star. The flight did much also for the reputation of the Moth, and was to lead to the boom in light aircraft construction in the United Kingdom. G-AAAH was presented to the Museum by the 'Daily Mail' in January 1931.

Focke-Achgelis Fa 330A-1 Bachstelze 100509

Built 1943, single seat observation rotor-kite.

This example of the Fa 330 is probably one of the best maintained examples left in the world. It is almost certainly an Fa 330A-1 variant. The A-1 was built by Weser Flugzeugwerke at Delmenhorst near Bremen. This example was one of many brought to the United Kingdom in 1945 and extensively test-flown at the Royal Aircraft Establishment, Farnborough, and the Airborne Forces Experimental Establishment, Beaulieu, towed behind vehicles. 100509 was transferred to the Museum from the Ministry of Supply in 1950.

Fokker E.III BAPC.56

Built 1915, single seat fighter, powered by one Oberusal rotary piston engine of 120 h.p.

This aircraft, believed to have been serialled 210/15, was presented to the Museum in 1918, having been captured as War booty. It is displayed inside the Museum as an uncovered frame to reveal its construction techniques. Nothing else is known of it.

Gloster E.28/39 W4041/G

Built 1941, jet-propelled prototype aircraft, powered by one Whittle W.2/500 turbojet of 1,700 pounds static thrust.

This aircraft was the Allied Powers' first jet propelled aircraft. It was built to contract SB/3229 by Glosters, under the design leadership of George Carter. It was built in the Experimental Department at Gloster's Hucclecote facility. However bombing forced the completetion of the aircraft to be halted whilst it was moved to the premises of Regent Motors Limited in Cheltenham. On April 7th 1941 it started taxi trials at Hucclecote and succeeded in making some short 'hops'. It was moved to Cranwell for actual flight testing and first flew there at 0745 hours on May 15th powered by a Whittle W.1 engine of 860 pounds static thrust. Testing continued and the aircraft was rebuilt in 1942/3 with a new wing and a Whittle W.1A engine of increased thrust and reliability, and flew again in this guise on on February 4th 1943. In May 1943 it was finally fitted with a Whittle W.2/500. It was used by the Royal Aircraft Establishment at Farnborough from April 1944 until retirement in 1946 and presentation to the Science Museum. It was placed on display inside the Museum on April 28th 1946.

Hawker Hurricane I L1592 'KW-Z'

Built 1938, single seat fighter, powered by one Rolls-Royce Merlin piston engine.

This aircraft was one of the first batch of fifty aircraft to be ordered into production. It only saw limited service with the Royal Air Force, serving initially with No.56 Squadron. Its second, and last, unit was No.615 (County of Surrey) Squadron, with whom it received the code letters 'KW-Z'. It was forced down by a Messerschmitt Bf 109 and crash landed near Croydon on August 18th 1940, and is thought not to have flown again. It joined the Museum in 1954.

JAP/Harding Monoplane BAPC.54

Built 1910, single seat monoplane, powered by one J.A.P. modified Anzani piston engine of 45 horse power.

This aircraft was based on the successful Bleriot Type XI monoplane and was made to the order of Mr. H. J. Harding by J. A. Prestwich (hence the designation of the type). It flew for the first time on April 10th 1910. With Mr. Harding piloting, it attended the Blackpool Aviation Meeting of 1911. It was stored throughout the First World War and beyond and presented to the Museum in 1930 by J. A. Prestwich Limited. It was allocated the identification number 54 by the British Aircraft Preservation Council.

Levavasseur-Antoinette Developed Type VII BAPC.55

Built 1910, single seat monoplane, powered by one Antoinette V-8 of 50 horse power.

This aircraft was built by Hubert Latham to standard Antoinette monoplane plans, using the improved Type VII for his pattern. This aircraft was to help Mr. Latham in his later experiments with monoplanes and rigid structure wings. It was presented to the Museum by Robert Blackburn in 1926. It has been allocated the British Aircraft Preservation Council identification number 55.

Lilienthal Standard

Built 1976, replica of the original single seat hang-glider.

This aircraft has been put on display to replace the original Lilienthal, bought by Percy Pilcher in 1895, as the original had become very fragile and is now in store.

Messerschmitt Me 163B-1 Komet 191316

Built 1943, single seater interceptor, powered by one Walter 109.509 bi-fuel rocket motor.

This aircraft was captured in Germany in 1945 and transferred to the United Kingdom. It more than likely was allocated an 'Air Ministry' number, but to date none has been traced for it. It was allocated to the Science Museum and was put on display after being reconditioned by the apprentices of No.1 School of Technical Training at Halton in 1961.

Pilcher Hawk replica BAPC.57

see photo 79

Built 1896, one man hang-glider replica.

This is a replica of Percy Pilcher's last hang-glider, the original can be found in the Royal Scottish Museum (q.v.). This replica has been allocated the British Aircraft Preservation Council identity number 57.

Roe Triplane No.1 BAPC.50

Built 1909, single seat triplane, powered by one JAP piston engine of 9 horse power.

This aircraft was A. V. Roe's second triplane and was the first ever machine to make a powered flight in the United Kingdom, that was of totally British construction. It was a very humble affair being covered with the legendary brown paper for its early flights. It flew for the first time on July 13th 1909 from Lea Marshes. It was not until 1925 that this important exhibit was obtained by the Museum. It has been allocated the British Aircraft Preservation Council identification number 50.

Rolls-Royce Thrust Measuring Rig XJ314

see photo 41

Built 1953, single seat vertical take off research hovering rig, powered by two Rolls-Royce Nene 101-IV turbojets, venting by direct efflux and through control jets.

This thrust measuring rig (or 'Flying Bedstead' from its looks akin to an airborne bed) was one of two designed and built by Rolls-Royce to aid its investigation into the development of jet engines for vertical take off. It was rolled out at Hucknall on July 3rd 1953. It made its first flight, tethered to the ground, six days later. XJ314 graduated to free flight on August 3rd. It was operated by Rolls-Royce until it was delivered to the Royal Aircraft Establishment on January 13th 1957. Up to this time it had made some 224 tethered flights, totalling nine hours flying time, and sixteen free flights, a free flight total of one hour 45 minutes. It suffered an accident with the RAE on September 16th 1957 and it was decided not to restore it to flying condition. Instead it was returned to Hucknall and rebuilt by Rolls-Royce to static exhibition standards and handed over to the Science Museum.

Royal Aircraft Factory SE.5A F939

Built 1918, constructor's number 688, single seat fighter, powered by one Wolseley Viper piston engine of 200 horse power.

Built in 1918 on contract to the Royal Air Force as F939 by Wolseley Motors at Adderley Park in Birmingham, and was delivered directly into long term storage. It was registered as G-EBIB to Major J. C. Savage in December 1924. It was nominally based at Hendon, but was to be seen all over the country, the Continent, and in the United States with Major Savage's Skywriting Company. It was fitted with modified exhausts to make smoke in order to write slogans for famous products in the sky. It was cancelled as withdrawn from use on August 6th 1935. It was presented to the Museum by Major Savage in 1939. It has been repainted in typical wartime colours.

Saunders-Roe Skeeter AOP. Mk.12 XN344

Built 1960, constructor's number S2/7150, aerial observation post, powered by one De Havilland Gipsy Major 215 of 215 horse power.

XN344 was awaiting collection by the Army Air Corps at Eastleigh on March 1st 1960, and was delivered to No.15 Maintenance Unit at Wroughton eight days later. It was despatched to Middle Wallop on October 11th 1960 and issued to No.652 Squadron seven days later. No.652 Squadron was based in West Germany, as was its next mount, No.654 Squadron, to whom it was officially issued on April 1st 1962. It was back at No.15 MU on November 23rd 1967 and was declared a non-effective airframe on May 27th 1968. The Army Air Corps had not finished with it however, it was allocated 8018M and moved to Middle Wallop on July 10th 1968 for instructional use. It was presented to the Museum via the Ministry of Defence in 1975.

Supermarine S.6B S1595

see photo 77

Built 1931, high performance racing floatplane, powered by one Rolls-Royce 'R' piston engine of 2,330 horse power.

This aircraft was built for the Royal Air Force High Speed Flight for the 1931 Schnieder Trophy Race. It was built at Southampton. In the race it was flown by Flight Lieutenant John Boothman on September 13th and with a speed of 340 miles per hour won the Trophy outright for the United Kingdom, as the rules of the Trophy were that if it were ever won three years on the run by one country, that country could maintain posession of the Trophy. It later went on to beat the World Speed Record at 407 mph. It joined the Museum in 1932.

Supermarine Spitfire IA P9444 'RN-A'

Built 1940, single seat fighter, powered by one Rolls-Royce Merlin II of 1,030 horse power.

P9444 was first delivered to the Royal Aircraft Establishment at Farnborough on April 5th 1940, but was at No.6 Maintenance Unit at Brize Norton on May 23rd. On June 4th it joined No.72 Squadron, receiving the code letters 'RN-A'. On July 3rd it crash landed heavilyand was removed to the care of No.1 Civilian Repair Unit on July 12th for debending. It was issued to No.45 MU at Kinloss on March 14th 1941. It re-entered active service on July 4th that year when it joined No.58 Operational Training Unit only once again to be hurt in an accident. It was not as serious as its first and was repaired on site from 19th to 23rd August 1941, going back into service with No.58 OTU the same day. However its luck was not in and it was declared Category Ac (beyond the capability of local repair) on December 17th. It was repaired on site by the 20th and rejoined the OTU. It fell foul of the ground again on May 5th 1942 when it was pronounced to have suffered a Category B (beyond repair on site) accident. It was ready for collection, repaired again on June 13th. By the 18th it was installed in No.8 MU at Little Rissington, pending issue to No.61 OTU on August 5th. It managed to see 1942 out without another accident, but was not long into 1943 (January 30th) before being declared Category B again. It was removed and repaired by Air Service Training at Hamble from February 7th until March 29th. On March 31st it arrived at No.12 MU at Kirkbride and was issued to No.53 OTU on May 12th. The old story took only forty one days to occur when R9444 was once again Category B. This time Scottish Aviation were entrusted to rebuild it, taking from November 14th to February 18th 1944. There then followed a succession of MUs : No.39 MU at Colerne from February 29th 1944, No.82 MU at Lichfield from May 10th 1944 and No.52 MU Cardiff from August 25th 1944. On August 28th 1949 it was allocated for museum purposes. It appeared at the Horse Guards Parade Display of 1949. It was moved to No.58 MU Newark in September 1951. It was then allocated to the Science Museum and moved to their store at Sydenham on December 16th 1954. It was repainted in 1961 in No.72 Squadron colours and erected in the Museum in 1963.

Vickers FB.27 Vimy IV BAPC.51

see photo 76

Built 1918, constructor's number possibly 13, converted bomber, powered by two Rolls-Royce Eagle VIII piston engines of 360 horse power each.

This aircraft was the thirteenth Vimy produced by Vickers at Weybridge and was earmarked for the use of Captain John Alcock (pilot) and Lieutenant Arthur Whitten Brown (navigator) who were to make an attempt on the prize of £10,000 offered by the 'Daily Mail' for the first non-stop trans-Atlantic flight. The Vimy (which may have been intended to be F8608 for the Royal Air Force) never carried an external identity and was test flown on April 18th 1919. It was then crated and shipped to Newfoundland, arriving on May 26th. It was then assembled by a Vickers team at Quidi Vidi. From Quidi Vidi it was flown to a place called Lester's Field on June 9th, from where the attempt was to be made. The historic flight took off on the 14th and ended in a crash ; in a hard landing in a bog at Clifden in Ireland sixteen hours 28 minutes later on the 15th. At an average airspeed of 118 miles per hour Alcock and Brown had flown 1,890 miles in terrible weather conditions. Both received knighthoods for their exploits. Vickers presented the machine to the nation on December 15th and it was used to inaugurate the aeronautics section of the Science Museum. Prior to opening the new gallery in 1963 the Vimy was totally recovered in fabric. A mock-up of the Vimy cockpit is also available for inspection.

Westland/Hill Pterodactyl 1 J8067

see photo 75

Built 1928, experimental tail-less aircraft, powered by one Bristol Cherub III piston engine of 32 horse power, driving a pusher propeller.

This experimental aircraft was built to test the theories of Professor G. T. R. Hill on the 'pterodactyl' wing planform which required no tail surfaces as such. This machine was by Professor Hill and his wife and was to be employed principally at Farnborough on aerdynamic work. Later models were produced in association with Westlands, and carried the Westland/Hill tag. This aircraft is strictly a Hill Pterodactyl. It was withdrawn with the advent of the side-by-side two seat versions. It was stored until presented to the Museum by Professor Hill in 1951.

Wright Flyer replica BAPC.53

Built 1948, static replica.

As mentioned in the introductory narrative, this aircraft was built at Hatfield by the De Havilland apprentices to replace the original Wright Flyer which returned to the United States for installation in the Smithsonian Institution in Washington. It is a faithful replica, but is not powered. It has been allocated the BAPC identification number 53.

Yokosuka MXY-7 Ohka II BAPC.58

Built 1944, rocket powered human-piloted flying bomb.

This aircraft was captured in 1945 from the Japanese and brought back to the United Kingdom as war booty and assigned to the Science Museum. It carries the 'serial' 15-1585. It was later allocated the British Aircraft Preservation Council number 58.

Notes : The following nose sections are also on display :-

Beagle B.206C Series 1 G-ATDD c/no. B.013, production number C.4, first flown May 12th 1965. Damaged at Sherburn June 27th 1973, broken up at Yeadon and nose to Museum in 1975.
De Havilland DH.106 Comet 1A G-ANAV c/no. 06013, first flown January 18th 1955, originally CF-CUM. Used for flight trials at Farnborough, broken up there 1955.
Douglas Dakota IV KN448 ex Royal Canadian Air Force, swopped for Curtis Seagull. Installed in the gallery modernisation of 1976.
English Electric Canberra B. Mk.5 VX185 Aircraft used for the first double crossing of the Atlantic. Rest of airframe later allocated 7631M.
Gloster Meteor III EE416 Donated by Martin Baker Aircraft, first aircraft to be used for live ejector seat trials.

Hawker Hunter F. Mk.6 Acquired from the Ministry of Defence in 1973.

The following balloon baskets or gondolas are on display :-

Royal Aircraft Factory Beta Airship control car, built 1910, presented to Museum 1923.
Cameron A-375 G-BBGN gondola, c/no. 90, World hot air balloon altitude record of 13,971 metres, established on January 25th 1974.
Piccard OO-BFH World gas balloon record altitude of 10½ miles, set in 1932.
Piccard O-56 G-ATTN c/no. 1352, 'Red Dragon', hot air ballon, first flown in 1966.
Short Brothers Balloon basket, of about 1900.

The following aircraft are owned by the Museum, but are displayed or kept elsewhere;

BAC/Aerospatiale Concorde 002 G-BSST displayed at the Fleet Air Arm Museum, Yeovilton.
Cierva C.24 G-ABLM in store at Hatfield for the Mosquito Aircraft Museum.
De Havilland DH.100 Sea Vampire I LZ551/G displayed at the Fleet Air Arm Museum, Yeovilton.
Weir W-2 displayed at the Museum of Flight, East Fortune.

The following aircraft are in store for the Museum, mostly at Hayes :-

Clarke Glider BAPC.100 biplane glider built for the Royal Aero Club in 1910 by T. W. K. Clarke and Company of Kingston-on-Thames.
Handley Page HP.39 Gugnunc G-AACN experimental biplane with full span slots and flaps, built in 1929, used by the Royal Aircraft Establishment as K1908. *see photo 88*
Lilienthal Glider replica BAPC.52 the original aircraft, in store.
Mignet HM.14 Pou du Ciel G-AEHM built by H. J. Dolman in 1936, c/no. HJD-1, in Bristol. Offered to the Museum in 1938.
Short SC.1 XG900 c/no. SH.1814, VTOL research aircraft, first flown February 4th 1957. Used by the RAE at Thurleigh until presented to the Museum on June 22nd 1971. *see photo 80*

SHUTTLEWORTH COLLECTION

Old Warden Aerodrome, in a pleasant rural area near Biggleswade, Bedfordshire. (Access via the A1. Trains to Biggleswade or Bedford, with the 179 United Counties bus service.)

In the later 1920s Richard Shuttleworth, the son of a rich land-owning family, started a collection of early road vehicles, extending his interest afterwards to include historic aeroplanes. By the outbreak of the War in 1939 he had obtained the 1909 Bleriot, the 1910 Deperdussin, the 1912 Blackburn and the 1916 Sopwith Pup. The first aircraft that he acquired in 1931 was a 1928 De Havilland Moth which he used as a personal fly-about. With the passage of time this became historic and has now lived on the same aerodrome for longer than any other aeroplane in existence.

Richard Shuttleworth was killed in a flying accident whilst serving in the Royal Air Force during World War Two, but after the War a charitable Trust was formed to continue the work that the founder had started. One aim throughout was that each exhibit must not be just a static museum piece, but that it must be fully workable. This policy is maintained today and the Shuttleworth Collection houses Europe's only full range of flyable historic aircraft, spanning the development of the flying machine from the earliest days.

The exhibits may be seen between 1000 and 1700 hours every day throughout the year except at Christmas; on special occasions (see below for the 1977 season dates) some of the historic machines are demonstrated in the air.

1977 flying days are as follows (full details will be sent on receipt of an S.A.E.) : April 10th/11th; April 24th Auster Pilot Club Fly-in; May 1st Kite Day; May 22nd tail-down light aircraft fly-in; May 29th; June 12th Aeromodeller Day; June 18th Air Squadron Aerobatic Trophy; June 26th; July 17th Shuttleworth Veteran Aeroplane Society - Model aircraft; July 31st Military Air Day; August 7th Performance Kits; August 28th/29th; September 25th; October 30th. Please note, these dates are published as a guide only, and the organisers reserve the right to alter dates and content.

There is a gift shop, restaurant and snack counter. The supporting Shuttleworth Veteran Aeroplane Society is the largest body of its kind and membership is open to all who wish to help the Collection's unique aims.

Entry prices for 1977 : Normal Days ; adult 50p, children 25p, free parking.
Flying Days ; adult £1, children 50p, car £1.
Coach passenger, cyclist or pedestrian £1, children 50p.

Address : The Shuttleworth Collection, Old Warden Aerodrome, Biggleswade, Bedfordshire, SG18 9ER.

Telephone : 076 727 288.

Auster B5 AOP. Mk.9 XR241

Built 1960, constructor's number AUS/B5/173 (unconfirmed), aerial observation post, powered by one Blackburn Bombadier in-line piston engine of 180 horse power.

The early history of XR241 is unknown, its first positive appearance in history is being ferried from Middle Wallop to St. Athan, via Blackbushe, for storage on April 18th 1967. In 1968 and 1969 it was on strength with the Advanced Fixed Wing Flight of the Army Air Corps at Middle Wallop. It was chosen in 1969 to be the mount of Major Mike Somerton-Rayner, A.A.C., to represent the Corps in the 'Daily Mail' England to Australia Air Race of December 1969. For this race the little Auster was allocated the civilian registration G-AXRR. After the race, the machine was presented in airworthy condition to the Trust as a gift in January 1970. It has remained airworthy ever since.

Avro 504K E3404

Built 1920, basic trainer, powered by one Le Rhone rotary piston engine of 110 horse power.

This aircraft started out life as a post-War Avro 504N basic trainer with an Armstrong Siddeley Mongoose radial engine. It was donated by De Havilland's to the Trust and it was converted to 504K basis to make it more representative of the thousands of this version which flew sterling service during the First World War and did so much to start civilian flying again after the Armistice. To achieve this conversion a rotary Le Rhone was installed (by the original builders, Avro). This engine was originally in a French Hanriot HD.1 fighter. E3404 was finished in time to take part in the film 'Reach for the Sky' about the career of Douglas Bader. After this brief spell of stardom, the 504 came to Old Warden and has been a consistent performer ever since.

Avro 621 Tutor K3215

see front cover

Built 1931, basic trainer, powered by one Armstrong Siddeley Lynx radial of 215 horse power.

K3215 was another aircraft to appear in the film 'Reach for the Sky'. The Tutor was Avro's answer to the need to replace the Avro 504N. This machine first served with the Central Flying School, and later with No.1 Flying Training School, Nos.41 and 61 Operational Training Units. It was stored throughout the Second World War and then refurbished to serve in the film. In 1947 it was registered G-AHSA and continued to fly with several private owners, until becoming part of the Trust in 1959. It is still airworthy.

B. A. Swallow 2 G-AFCL

Built 1937, constructor's number 462, two seat light plane, powered by one Pobjoy Cataract III radial piston engine of 90 horse power.

G-AFCL was built at Hanworth by the British Aircraft Manufacturing Company and was registered on November 2nd 1937. It was based at Croydon in the years left before the War, with W. L. Hope and then C. McCarthy. It is thought to have been stored throughout the War, being restored to the register in October 1948 to G. H. Forsaith. It has spent most of its post-War life based at Old Warden, and it is currently based with the Trust, owned by A. M. Dowson.

Blackburn Monoplane No.9

Built 1912, works order number 725, single seat monoplane, powered by one Gnome rotary piston engine of 50 horse power.

This aircraft was built by Blackburns to the order of Mr. Cyril Foggin. It was sold in 1913 to Mr. Francis Glew who placed it into store at Wittering in 1914 at the start of World War One. It was discovered in its hiding place, in reasonable condition, in 1938. It was put back into storage however with the advent of the Second World War. It was restored to flying condition in 1946/7. It has been kept airworthy to date. It has been allocated the British Aircraft Preservation Council identification number 5.

Bleriot Type XI No.14

Built 1909, single seat monoplane, powered by one Anzani piston engine of 25 horse power.

This aircraft is thought to have been one from the Bleriot School at Hendon, opened in October 1910. It was bought by a Mr. Grimmer in damaged condition in 1912, he rebuilt it to flying condition and then taught himself to fly it, using a polo field in Ampthill for his aerodrome. When flying was forced to stop in 1914 he stored it in his home town and it was discovered in 1935 in a garage. Richard Shuttleworth restored it and it was flown in the Royal Air Force Display at Hendon in 1936. It has since been maintained in airworthy condition. It has been allocated the B.A.P.C. identification number 3.

75 *Pterodactyl J8067 in big company at RAF Pageant 1926* *'Flight' RAFD 1926/A*

76 *Moment of history, Alcock & Brown's Vimy takes off, 14 June 1919* *Science Museum 601/38*

77 *The Science Museum's record breaking S-6B S1595, on the water 1931* *'Flight' 11567*

78 *Original Flea G-AEKR, Finningley 1967* S. G. Jones

79 *Pilcher Hawk in flight 1896* Science Museum 275/52

80 *Short SC.1 XG900 on show at Farnborough* P. H. Butler

81 *Hurricane Z7015 at Old Warden* R. W. Cranham

82 *Tomtit G-AFTA at Baginton 19/6/54* R. A. Scholefield

83 *Magister G-AJDR at Speke 24/5/62* P. H. Butler

Bristol 96 F.2b Fighter D8096

Built 1918, constructor's number 7575 (unconfirmed), two seat fighter, powered by one Rolls-Royce Falcon piston engine of 275 horse power.

This 'Brisfit's' early history is not known, although it is most likely not to have seen much operational flying. It was bought in a disposal sale in 1936 by Mr. C. P. B. Ogilvie who intended converting it to civilian status (the Bristol Fighter was called the Bristol Tourer in its civil role) and reserved the registration G-AEPH for the purpose. Mr. Ogilvie did not carry through the conversion. Like several other Shuttleworth items, it was found much later, in 1949, in a shed, this time in Elstree. In 1950 it was moved to Filton where it was fully restored by the makers and Rolls-Royce. It has been one of the most popular members of the Trust's airworthy fleet ever since.

Cierva C.30A G-AHMJ

Built 1937, single seat autogiro, powered by one Armstrong Siddeley Genet Major radial piston engine of 140 horse power.

Built by Avro as a Type 671 Rota I for the Royal Air Force as K4235, this machine first entered service with the School of Army Co-operation at Odiham, followed by No.1448 Flight at Halton. No.1448 Flight was renumbered and up-graded to Squadron status, No.528, in June 1943. It continued to serve, principally in the radar calibration role, until being sold off for civilian use on May 23rd 1946. It was sold to the Fairey Aviation Company at White Waltham, and they registered it as G-AHMJ, as a Cierva C.30A, on July 11th. It is thought that Fairey's used G-AHMJ to train their rotary wing test pilots, as the company was gearing up to produce some very advanced rotorcraft. G-AHMJ was dismantled at White Waltham in July 1947 and donated to the Hayes and Harlington Sea Cadets, for use as an instructional airframe. (Fairey's main plant was at Hayes.) From here it was acquired by the Trust and is now nearing the end of a restoration that will bring it back to Rota I status as K4235.

De Havilland DH.51 G-EBIR

see photo 85

Built 1925, constructor's number 102, touring biplane, powered by one Airdisco Renault in-line piston engine of 120 horse power.

G-EBIR was built at Stag Lane and was delivered by air to Kenya in January 1926, to become the first aircraft on that country's civil register, as G-KAA, registered in September 1928. By January of the following year Kenya had been given a registration sequence in the colonial system and was re-registered as VP-KAA. By this time it had acquired the name 'Miss Kenya'. VP-KAA was to survive the rugged climate perfectly well, and in July 1965 it was flown home for the Shuttleworth Trust, this time not using its own power, but inside the cavernous hold of a Beverley freighter. After storage for some while G-EBIR was moved to Hawker Siddeley's plant at Hawarden and there a complete rebuild to flying condition was undertaken. 'Miss Kenya' flew again in March 1973 and joined the Trust soon afterwards.

De Havilland DH.53 Humming Bird G-EBHX

Built 1923, constructor's number 98, single seat ultra-light, powered by one A.B.C. Scorpion twin cylinder piston engine of 35 horse power.

G-EBHX was built in 1923 for the Lympne Light Aircraft Trials of the same year. It first flew in September. It was called 'Humming Bird' and the name stuck with the type. In 1925 it was re-engined with a Tomtit and re-named 'L'Oiseau Mouche'. It managed to survive the War stored in a shed at Betteshanger Colliery in Kent, and was found by the Trust. It was overhauled and restored to flying condition by the De Havilland Technical School and first flew again after rebuild on August 4th 1960. It continues to be airworthy.

De Havilland DH.60X Hermes Moth G-EBWD

Built 1928, constructor's number 552, touring biplane, powered by one Cirrus Hermes in-line piston engine of 105 horse power.

G-EBWD originally flew from Stag Lane with a Cirrus II engine, but was later brought up to DH.60X standards. It is typical of the Moth era of pre-War Britain and was bought by Richard Shuttleworth, who used it for his personal fly-about and, as stated in the introductory narrative, has the distinction of being based at one aerodrome for longer than any other aeroplane. Richard Shuttleworth bought G-EBWD in 1931 at Brooklands. It is still airworthy.

De Havilland DH.82A Tiger Moth G-ADGV

Built 1935, constructor's number 3340, basic trainer, powered by one De Havilland Gipsy Major in-line piston engine of 130 horse power.

G-ADGV was built at Stag Lane and served with Brooklands Aviation, and from 1939 with No.6 Elementary and Reserve Flying School which was run by Brooklands. On September 7th 1940 it was impressed and allocated the serial BB694, it stayed with No.6 E&RFS however, although the unit was now known as No.6 Elementary Flying Training School. On July 5th 1942 it moved to No.29 EFTS at Clyffe Pypard. On August 24th 1946 it was retired to No.5 Maintenance Unit at Kemble but was moved to Royal Naval Air Station Stretton on October 15th. Here it apparently stayed until being recorded at Rochester on April 8th 1955. By December 14th of that year it was at Lee-on-Solent, no doubt on tug and 'hack' duites. On July 13th 1956 it was posted to Lossiemouth, on September 22nd 1959 to Arbroath and it ended its Navy days with the Royal Naval Engineering College at Roborough from November 17th 1960. On September 8th 1972 it was restored as G-ADGV with Air Commodore A. H. Wheeler and is nearing the completion of its restoration with Shuttleworths and Bowker Air Services at Rush Green. It was allocated G-BACW when it was restored, but the mistake was soon noticed and this was not taken up. This is Allen Wheeler's private mount, but can effectively be considered as part of the Trust's fleet.

De Havilland DH.82A Tiger Moth T6818

Built 1932, constructor's number 85087, basic trainer, powered as G-ADGV above.

T6818 was built at Stag Lane for the Royal Air Force, its early career with them being unknown. We take up the story from October 1940, the 20th to be exact, when it was issued to No.21 Elementary Flying Training School at Booker. On September 26th 1942 it was moved to No.10 Maintenance Unit at Hullavington for storage until being issued after a considerable period of inactivity to the Station Flight of Aston Down on October 8th 1951. This was shorted lived however as T6818 was put back into store, this time at No.20 MU at Aston Down on March 30th 1953. On November 6th it was sold to A. J. Whittemore (Aeradio) Limited and registered to them as G-ANKT on December 24th. It was then to go through a succession of private owners until being registered to A. H. Wheeler, trading as the Shuttleworth Trust, on August 12th 1971. It has since been refurbished and put back into its military guise.

De Havilland DH.88 Comet G-ACSS

see photo 86

Built 1933, constructor's number 1999 (later changed to 2260), two seat high performance racer, powered by two De Havilland Gipsy Six R in-line piston engines of 230 h.p. each.

The beautiful DH.88 was born from a need to have an entry in the 1933 Robertson Race that would stand a chance against the American entrants (the Douglas DC-2 of KLM etc.). It was designed in utter secrecy with considerable speed. Sir Macpherson Robertson had put up a prize for the winning aircraft in an England-Australia (Melbourne) race, for all-comers. G-ACSS was built to the order of A. O. Edwards, the manager of the Grosvenor House Hotel. He had contracted with pilots C. W. A. Scott and Tom Campbell Scott to fly it for him. After only a brief development stage, G-ACSS, 'Grosvenor House' in red and ivory was lined up at the start at Mildenhall on October 20th 1934. The Comet won in the time of 70 hours, 54 minutes and was to earn an unforgettable place in aviation history (as, of course did the Grosvenor House Hotel, which was the main aim of the exercise). G-ACSS's history did not stop there, it was flown at Martlesham Heath in May 1935 for evaluation as K5084. Whilst appearing at the 1936 Hendon it crashed at nearby Batley on August 19th 1936. It was bought in this bent state by F. E. Tasker, who had the Essex Aero Club rebuild it for him at Gravesend. This time it was fitted with Gipsy Six IIs. In pale blue scheme G-ACSS became 'The Orphan'. Later as 'The Burberry' A. E. Clouston flew it out to the Cape of Good Hope and back again in the record time of 15 days 17 hours in November 1937. In 1938 it tried for another record run to Australia, this time named 'Australian Anniversary', but its undercarriage collapsed in Cyprus. It was repaired and made Sydney on March 15th in a time of 80 hours 56 minutes going on to Blenheim in New Zealand in a further 7½ hours. It was back at Croydon on March 26th! G-ACSS had travelled a total of 26,450 miles in a time of ten days, 21 hours, 22 minutes. That record still stands. G-ACSS then went into store at Gravesend, until emerging again before 1951. It was restored by the ever-eager De Havilland Technical School and was displayed at the Festival of Britain of 1951. On October 30th 1965 it was handed over to the Shuttleworth Trust. The Trust are painstakingly rebuilding G-ACSS so that it will fly again one day, this is now expected to take place in 1980. This is an exceptionally costly operation, and donations will be well received by the Trust at the address given in the introductory narrative.

Deperdussin Monoplane No.43

Built 1910, single seat monoplane, powered by one Anzani piston engine of 35 horse power.

Deperdussin No.43 was built in 1910 and was bought by Mr. Grimmer of Ampthill in 1914, see under the Bleriot for more information. It was discovered by Richard Shuttleworth in 1935, and restored to flying condition by 1937. It has been allocated the British Aircraft Preservation Council number 4.

Desoutter I G-AAPZ

see photo 52

Built 1931, constructor's number D.25, light cabin monoplane, powered by one Menasco Pirate piston engine of 130 horse power.

G-AAPZ was registered to National Flying Services Limited, based at Hanworth, on August 19th 1931. It was registered to W. S. Stephenson also at Hanworth in December 1932 before becoming the property of Richard Shuttleworth in February 1935. After a period of loan to the Torbay Aircraft Museum G-AAPZ is now back at Old Warden and is being restored to full flying condition in the hands of some private individuals.

English Electric Wren No.4

Rebuilt in 1957 using two airframes, single seat ultra-light, powered by one A.B.C. Scorpion piston engine of 3 horse power.

Built for the 1923 Lympne Light Aircraft Trials at Lytham St. Annes, Wren competition number 4 (constructor's number 3) came into the hands of the Trust at an early stage. In 1957 it was rebuilt at Warton to flying condition using many components from Wren competition number 3 (constructor's number 4) donated by Mr. R. H. Grant of Dumfries. Number 3 had been registered in 1926 as G-EBNV, but No.4 never acquired such status. The new composite flew for the first time in January 1957. No.4 has been allocated the British Aircraft Preservation Council identity number 11. Considering its tiny powerplant, the Wren flies on regular occasions.

Gloster Gladiator I L8032

Built 1937, single seater fighter, powered by one Bristol Mercury VIIIA radial piston engine of 840 horse power.

Built by Glosters in 1937 as a Gladiator I the War-time history of this 'Glad' is unknown, although it is believed to have flown at one time with a Meteorological Reconnaissance Flight. It was bought back by the makers in February 1948 (along with Sea Gladiator N5903) and was stored until November 1950 when it was moved to Air Service Training at Hamble as an instructional airframe (N5903 went to Ansty). In 1951 it was bought by Viv H. Bellamy who overhauled it at Eastleigh and registered it as G-AMRK. He flew it regularly until it was again bought by Glosters in August 1953, and G-AMRK continued to fly often. In 1956 it was repainted in military markings, but as K8032, and was eventually presented to the Trust on November 7th 1960. After a complete overhaul by Flight One Limited at Staverton in 1974 the Gladiator took on its real identity, L8032.

Granger Archaeopteryx G-ABXL

Built 1930, constructor's number 3A, tail-less single seater ultra-light, powered by one Bristol Cherub I of 32 horse power.

This parasol tail-less aircraft was designed by brothers R. F. T. and R. J. T. Granger, with some assistance from C. H. Latimer-Needham. It was built at Attenborough, and flew for the first time at Hucknall in October 1930. Only in 1932 was it declared to the authorities and registered as G-ABXL. It was placed into store through the War, at Chilwell, and afterwards presented to the Trust. It was refurbished and flew again in 1971.

Hawker Afghan Hind BAPC.78

Built circa 1937, two seat light bomber, powered by one Rolls-Royce Kestrel of 640 h.p.

This aircraft is one of two discovered in open store with the Afghanistan Air Force and brought back to the United Kingdom for preservation. The other is with the Royal Air Force Museum. This aircraft, allocated the British Aircraft Preservation Council indentity number 78 in the absence of any other identity, is being restored to flying condition at Old Warden. It wasbrought back from Afghanistan overland, with the assistance of the Ford Motor Company. Its history is as vague as its identity, details welcomed.

Hawker Tomtit K1786

see photos 82 & 84

Built 1929, basic trainer, powered by one Armstrong Siddeley Mongoose radial piston engine of 150 horse power.

K1786 was delivered to the Royal Air Force in January 1931 and entered service with No.3 Flying Training School, later with No.23 Group Communications Flight and then No.5 Group Communications Flight. It was withdrawn from use and sold off on August 26th 1938. The following year it was registered to the Leicestershire Aero Club as G-AFTA on April 28th. It moved on to several private owners, the most notable of which being Alex Henshaw from June 1942, and Neville Duke from April 1949. In July 1950 it was back in the hands of the original builders. It flew with Hawkers until September 1959 when it was donated to the Trust. In 1967 it was fully restored by Hawker Siddeley at Dunsfold and restored in its original military markings. It is still airworthy.

Luft Verkehrs Gesellschaft m.b.H LVG C.VI 7198/18

Built 1918, two seat fighter, powered by one Benz Bz IVa in-line piston engine of 230 h.p.

The German Air Force order number, 7198/18, that this aircraft now wears may or may not be authentic. Built in Berlin in 1918 the C.VI was acquired at the end of the First World War and taken to Martlesham Heath for evaluation. On August 27th 1919 it became one of the Imperial War Museum's charges. It was displayed at Crystal Palace from 1920. From March 24th it was moved to the Science Museum store and remained there until 1932, when it was moved to Cardington. In 1936 it moved to Hendon where it was restored by Flight Lieutenant Buckle and Flight Sargent Scott to full flying condition. It flew at the RAF Display at Hendon on June 26th 1937. After this brief spell in the air, it was returned to Cardington for storage until moving on to No.5 Maintenance Unit at Kemble in 1945. It was displayed at Kemble that December , but only as a rather forlorn fuselage. It moved on to Fulbeck and then to No.39 MU at Colerne in 1950. It was prepared here for exhibition at the RAF Display at Farnborough that year. By 1955 it was in store at No.2 MU Stanmore Park and was noted wearing the number '1594' in September that year. In 1959 it was loaned from the Air Historical Branch to the Shuttleworth Trust. It first flew after restoration at Old Warden on September 28th 1972.

Manning-Flanders MF.1 G-BAAF

Built 1973, constructor's number PPS/REP/8, single seat monoplane, powered by a Continental A75 flat four piston engine.

Before the First World War W. O. Manning and L. H. Flanders designed their MF.1 monoplane. Nothing came of the design until Doug Bianchi and his Personal Plane Services decided to build what was effectively the prototype in 1973, following as closely as possible the original, but using a modern powerplant. It first flew from Booker on March 1st 1974. It was loaned to the Trust in late 1975.

Mignet HM.14 Pou du Ciel G-AEBB

Built 1936, constructor's number KWO.1, single seat ultra-light, powered by one Scott A.2S 'Flying Squirrel' twin cylinder piston engine of 25 horse power.

G-AEBB was built in January and February 1936 by Kenneth W. Owen in Southampton. It was first fitted with a Henderson 1300 c.c. motor cycle engine and is thought to have flown. It was stored throughout the War's early years until being donated in 1941 to No.424 Squadron Air Training Corps (see also under the R. J. Mitchell Memorial Hall) for use as a training airframe. It was kept by the unit until the early 1960s when it was donated to the Trust. The Trust fitted the more appropriate Scott engine, as used in many 'Fleas', and it taxied again for the first time on September 8th 1968. Although fully airworthy G-AEBB is not flown as this type of 'Flea' has been proved to be unstable in certain flight regimes.

Miles M.2L Hawk Speed Six G-ADGP

Built 1935, constructor's number 160, single seat racing aircraft, powered by one De Havilland Gipsy Six 1F in-line piston engine of 200 horse power.

G-ADGP was built at Woodley for Luis Fontes, and was registered on June 13th 1935. It was bought by Miles Aircraft Limited post-War and registered to them in August 1946. It was bought then by R. R. Paine in February 1948 and modified extensively, including a bubble canopy for racing. It was to compete in the annual Kings Cup Race from 1949. This aircraft still holds the Class C.16 closed 100 kilometers circuit record of 192·83 m.p.h. It was bought by David Hood and rebuilt at Castle Donington in 1971 and piloted by Mr. Paine came second in the 1972 Kings Cup. G-ADGP is still owned by Mr. Hood and lives with the

Trust at Old Warden.

Miles M.14A Magister P6382

see photo 83

Originally built 1940, constructor's number 2169 (but see below), primary trainer, one De Havilland Gipsy Major in-line piston engine of 130 horse power.

The history of this aircraft is exceptionally complex, being a very good example of the agonies an aviation historian may have to go through to find the real aircraft behind an identity! This aircraft delivered to the Royal Air Force at No.46 Maintenance Unit at Lossiemouth on October 29th 1940. It was issued to the Headquarters, Air Transport Auxiliary at White Waltham on April 10th the following year. It was damaged in a Category B (beyond repair on site) accident on January 29th 1943, being taken to No.51 MU at Lichfield on Apri 11th. It was declared surplus post-War and bought by Bristol Gliding Club. It was registered as G-AJDR on January 27th 1947. From May 1947 it was with Weston Aero Club, and with Gloucester Flying Club from June. In 1948, with its Certificate of Airworthiness lapsed, it was sold as scrap and fetched £8. It was rescued by C. J. de Vere in May 1950 and moved to Blackbushe. Here its identity began to be eroded. It was given the wings of G-AKMU. It moved to Redhill, but came to grief in a test flight in 1951, and was stored until 1953. In March of that year it was registered to the Redhill Flying Club. They used the fuselage of G-AJRS in an extensive rebuild. Also employed was the centre section, ailerons, starboard wing and tailplane of G-AJRU. To add to this, the port wing came from G-AJRT. To avoid any confusion the original G-AJDR's fuselage was burnt in May 1954. The new aircraft, officially G-AJDR was sold to W. S. Shackleton and on to the Scottish Flying Club at Renfrew in July 1954. By November 1957 it was flogging the circuit with the Denham Flying Club until damaged on May 10th 1959. Now the wings of G-AIDF were fitted. In August the port wing of G-AJZH was fitted. G-AJDR was based at Speke during 1960 and 1961 with C. Harrop. In April 1963 'G-AJDR' could be found at Christchurch. By September 1969 it was in the hands of Shoreham Aviation Services. On April 3rd 1970 it was sold to the Shuttleworth Trust and underwent a major rebuild. It emerged from this in superb War-time markings with the serial P6382 - this coming from the rear fuselage of G-AJRS and no doubt discovered still present under many coats of dope.

Miles Boxkite replica BAPC.2

Built 1964, constructor's number BM.7279, flying replica for film use, powered by one Rolls-Royce/Continental C90 flat four piston engine of 90 horse power.

This aircraft was built to order for the film 'Those Magnificent Men in their Flying Machines' and portrayed the American entry 'The Phoenix Flyer'. Although said in the film to be a Curtiss it was in fact based upon the Bristol Boxkite. It was built by F. G. Miles Limited at Ford and, being the first of three ordered, was known affectionately as 'Box 1'. It first flew under the power of a Rolls-Royce/Continental A65 of 65 horse power in April 1964. This engine gave poor results (despite being over three times the power of the original Boxkite's powerplant) and the C90 had to be installed. After the film had been completed it was returned to Ford for refurbishing and it was presented to the Trust, and has flown consistently since. It has been allocated the British Aircraft Preservation Council identification number 2.

Parnall Elf II G-AAIN

Built 1929, constructor's number 2 (later changed to J6), two seat biplane, powered by one ADC Cirrus Hermes in-line piston engine of 105 horse power.

This aircraft was one of three built by the Parnall Company at Yate. Its history was none too eventful, being stored through the War and beyond until becoming part of the Trust in the mid-1960s. It had been on loan to the Historic Aircraft Museum at Rochford for a while from its opening, but is now back at Old Warden and undergoing a full rebuild to flying condition.

Percival Gull Six G-ADPR

Built 1932, constructor's number D.55, light cabin monoplane, powered by one De Havilland Gipsy Six of 185 horse power.

This aircraft achieved fame as the aircraft in which Miss Jean Batten flew to Brazil in in 1935. It was later used for other epic solo flights to Darwin in five days, 21 hours and to New Zealand in 11 days 45 minutes. It was impressed for War service on July 17th 1940. It was delivered to Ringway on the 24th and used by the Station Flight as AX866. It was transferred to No.4 Glider Training School on April 8th 1942 and stayed with the unit (later renamed No.2 (Pilot) Advanced Flying Unit) until it was ferried to No.5

Maintenance Unit at Kemble on April 20th 1945 for disposal. It was bought by its builders in January 1946 and restored as G-ADPR. 'Jean' as the Gull was called was stored at Luton from 1959 until handed over to the Trust on April 25th 1961 by the company that had taken over Percivals, Huntings. G-ADPR was flown to Old Warden that day.

Percival P.56 Provost T. Mk.1 XF836 'J-G'

Built 1955, constructor's number P.56/339, advanced trainer, powered by one Alvis Leonides radial piston engine of 550 horse power.

XF836 was ready for collection by the Royal Air Force at Luton on June 17th 1955, and was issued to No.12 Maintenance Unit at Kirkbride five days later. It was first issued on a temporary basis to Manchester University Air Squadron for use on summer camp from April 29th 1957 until being issued to the Royal Air Force College at Cranwell on August 30th. It was coded 'J-G' at Cranwell and it wears these marks again today. It was delivered to No.49 MU at Colerne on July 11th 1960. It was next issued to the Central Navigation and Control School at Shawbury on July 26th 1960, coded 'R'. The unit was later called the Civil Air Traffic Control School. On January 6th 1969 it was transferred to the local No. 27 MU and was sold to the Trust on April 18th 1969. This Provost still flies in its military markings by special arrangement.

Roe Triplane Type IV replica BAPC.1

Built 1964, flying replica for film use, powered by one De Havilland Cirrus II in-line piston engine of 80 horse power.

Like the Boxkite above this was a replica for the film 'Those Magnificent Men in their Flying Machines' and was built by the Hampshire Aero Club at Eastleigh under the design leadership of Ray Hilbourne. It was based as accurately as possible on A. V. Roe's pioneering triplane of 1910, but employed modern techniques where-ever these would not deter from its external looks. After the film it was presented to the Trust, and it has remained airworthy ever since. It has been allocated the British Aircraft Preservation Council identification number 1.

Slingsby T.57 Sopwith Camel replica C1701

Built 1969, constructor's number 1701, flying replica for film use, powered by one Warner Super Scarab radial piston engine.

'C1701' was built to order by Slingsby Aircraft Limited at Kirbymoorside for use in a film that in the end was not made. It was registered as G-AWYY and first flew at White Waltham on March 4th 1969. It was sold in the United States in about 1970 but returned, to Old Warden in a crate in 1976. It has since been restored and is now flying with the Trust on loan from Mr. A. Carlisle. It is hoped to re-engine the replica with some form of rotary in the future.

Sopwith Pup N5180

Built 1918, single seat fighter, powered by one Le Rhone rotary piston engine of 80 h.p.

This aircraft was assembled from unused Pup parts as the civil variant, the Sopwith Dove. It was registered in 1925 to D. L. Hollis Williams as G-EBKY and was withdrawn from use in December 1933. The Dove was not much of a success, although it maintained the handling qualities of the fighter, it also was not an economic aircraft to operate. It was taken on by Richard Shuttleworth in July 1937 and reconverted to Pup status. It first flew in this guise on February 26th 1938, it was originally marked as N5184, but was altered to carry N5180 which is believed to be its correct previous identity. The Pup is maintained in airworthy condition and is one of the most called-for of the Trust's aircraft.

Southampton University Man Powered Aircraft BAPC.7

Built 1961, single seat man-powered aircraft.

This aircraft has a name almost as unwieldy as its huge dimensions, so is most often referred to as 'SUMPAC'. Design of this aircraft started in July 1960, with construction getting underway in January of the following year, being completed by September. It had the distinction of having made the first true man powered flight in the United Kingdom at Lasham on November 9th 1961, when it achieved a distance of 50 yards and a height at the maximum of six feet. In 1964 it was refined, but was damaged and donated to the Trust.

Southern Martlet G-AAYX

Built 1931, constructor's number 202, single seat sporting biplane, powered by one De Havilland Gipsy II in-line piston engine of 120 horse power.

Designed by F. G. Miles as a single seater aerobatic sporting aircraft based on the Avro Baby, G-AAYX was built in 1931. It spent most of its life with Butlins Limited, being stored throughout the War. It was overhauled by Miles Aircraft Limited and flew again after the War on June 17th 1947 and was redelivered to its base at Pwllheli. It became part of the Trust in 1955, and is currently stored having been partly overhauled by Hawker Siddeley Aviation at Woodford in the early 1970s.

Notes : All the above aircraft are generally available for public inspection at Old Warden. Although some may be inside the workshop. The following is a list of aircraft in store at Old Warden and not viewable by the general public.

ANEC II G-EBJO built 1924 for the Lympne Trials, c/no. 2, competition number 7, powered by one Bristol Cherub of 32 h.p. Presented to the Trust in 1937.
Blake Bluetit BAPC.37 composite aircraft composed of Spartans G-AAGN and G-AAJB and some Avro 504 parts. Built by W. H. C. Blake and flown in the mid-1930s.
Bristol 96 F.2b Fighter fuselage frame only, from Weston-on-the-Green.
De Havilland DH.82B Queen Bee LF858 1940 pilotless drone version of the Tiger Moth, remains only.
De Havilland Canada DHC-1 Chipmunk G-AOTD c/no. C1/0040, ex WB588, awaiting restoration, so far unconverted from military standard, from Farnborough.
Dixon Ornithopter BAPC.8 little known of type, in store for a long time.
EoN Primary BAPC.35 glider, c/no. EoN/P/063, previously on loan to the Torbay Aircraft Museum, called 'Verna Flo'.

The following aircraft are owned by the Trust but are elsewhere for various reasons

Royal Aircraft Factory SE.5A F904 ex G-EBIA, kept at Farnborough.
Avro 652A Anson C. Mk.19 Phase 2 TX183 under restoration to flying condition at Duxford.
BAT FK.23 Bantam G-EACN c/no. 15 ex K123, single seat biplane, under restoration at Shipdham. Long term project.
De Havilland DH.87B Hornet Moth G-ADND c/no. 8097, impressed as W9385, airworthy and active with the Hawker Siddeley Flying Club at Hawarden, on loan from the Trust.
Hafner R.II Revoplane BAPC.10 on loan to the Torbay Aircraft Museum.
Hawker Sea Hurricane I Z7015 fuselage etc. under restoration with Dowty Rotol at Staverton. Wings under restoration with British Airways at Heathrow.
Hunting P.84 Jet Provost 1 G-AOBU c/no. P.84/6, ex G-42-1, on loan to Loughborough College of Technology as an instructional airframe.
Supermarine Spitfire V AR501 airworthy and based at Duxford.
Supermarine Spitfire PR.XI PL983 under restoration to flying condition at Duxford.

SKYFAME AIRCRAFT MUSEUM

Skyfame Aircraft Museum is situated at Staverton Airport, just off the A40/B4063 highway between Gloucester and Cheltenham. It is open every day, except Christmas and Boxing Day, and the admission charge is 50p for adults and 25p for anyone of school or college age. There are special rates for parties and conducted tours.

Skyfame was the first aircraft museum in Great Britain, opening on August 31st 1963, with the main object of saving certain Second World War British aircraft, then vanishing at an alarming rate. The Museum actually owed its birth to the successful campaign organised by its founder, Peter Thomas, to save the Sunderland flying boat ML824 which had been handed over by the French Navy to his Short Sunderland Trust at Pembroke Dock in 1961. After eight years on view at the dock, the Trust handed the Sunderland over to the Royal Air Force Museum, where it is now on display at Hendon.

Skyfame's sub-title 'Desmond Commonwealth Flying Memorial' indicates that it is dedicated to the British Commonwealth Air Forces men and women of all races in the Second World War who, like Desmond Thomas, a bomber pilot flying Wellingtons, gave their lives in the six years of conflict. There are aircraft in the Museum which were used by all the Commonwealth Air Forces in this period and Skyfame is almost unique in exhibiting aircraft that served with every command of the Royal Air Force and the Fleet Air Arm and Army during the Second World War. There is the Tempest fighter which was our main defence against the flying bombs and the (then) new and fast German jet aircraft. The Mosquito bomber that did most of the work in the film '633 Squadron'. The Anson Mark I anti-submarine bomber represents the type that holds the record for RAF service - 34 years. The only flyable Airspeed Oxford trainer that trained more RAF pilots than any other type. The Saunders Roe A.1 prototype which was the world's first jet flying boat. A thirty ton Hastings four engined transport which booked parties enjoy entering to inspect, a diminutive home-made Flying Flea, a Firefly Royal Navy strike fighter and many other aircraft.

The model aircraft collection is the largest on permanent view in the country, numbering upwards of a thousand and includes the oldest model aircraft in Britain, the Cody 'Cathedral'. There is also a fine collection of pictures, aero-engines and many other interesting exhibits.

The Museum holds an annual air display which is the main west of the Midlands event of its kind, held with RAF participation and co-operation. Free entry to this is one of the many privileges enjoyed by members of the Skyfame Supporters Society, about which details can be obtained from the Museum.

Address : Skyfame Aircraft Museum, Staverton Airport, Cheltenham, Gloucestershire.

Airspeed AS.40 Oxford I V3388

Built 1940, constructor's number 3083, multi-purpose crew and pilot trainer, powered by two Armstrong Siddeley Cheetah 10 radial piston engines of 375 horse power each.

This Oxford was built by De Havilland Aircraft at Hatfield as V3388. It was civilianised in 1946 for use by Boulton Paul Aircraft Limited. It was registered as G-AHTW to that company on June 19th 1946. It was to serve the company faithfully in the communications role for fourteen years, being finally with-drawn from their service in December 1960. By this time it had appeared at Elstree and it was here that it was acquired by the Museum in February 1960, it was painted up as its former military self and in typical camouflage. It was delivered to Staverton on March 25th of that year, and has been with the Museum in flying trim since that date.

Auster B5 AOP. Mk.9 XR267

Built 1965, aerial observation post, powered by one Blackburn Bombadier 203 in-line piston engine of 173 horse power.

XR267 was issued to No.19 Maintenance Unit at St. Athan on June 3rd 1966. It is thought that the craft had little, if any, service use with the Army Air Corps, as it was declared a non effective airframe at St. Athan on November 8th 1968. On October 15th the following year it was bought by Mr. 'Dizzie' Addicott and moved to Holder's Garage in Congresbury, until sold to Toogoods Caravans, who had a garage up on Bedminster Down near Bristol, in 1973 for display purposes. XR267 was mounted at the garage and in order to attract custom was painted in a very gaudy scheme. It was acquired by Skyfame in 1974 and moved by road to Staverton on November 17th. It has since been returned to a more fitting colour scheme.

Avro 652A Anson I N4877 'VX-F'

see photo 89

Built 1938, maritime recconnaissance bomber, powered by two Armstrong Siddeley Cheetah IX radial piston engines of 350 horse power each.

N4877 was taken on charge by the Royal Air Force in 1938 and it is known to have served with No.2 Ferry Pilots Pool based at Filton and later with the Headquarters Air Transport Auxiliary at White Waltham. In both cases the aircraft was used for the role that the 'Annie' was to adopt for a huge span of years, that of communications, other than its original role of anti-shipping and anti-submarine work. It was struck off RAF charge in 1950 and was converted to civil status by R. L. Whyham at a leisurely pace at Squires Gate. It was given the registration G-AMDA on October 10th 1953. It was sold to Derby Aviation on completion and then joined the London School of Flying at Elstree, who used it for charters and other duties. In 1963 it was acquired by Skyfame, who repainted it in wartime colours of an Anson of No.206 Squadron, coded 'VX-F'. It was airworthy with the Museum until 1972.

Bristol B.171 Sycamore III G-48/1

see photo 1

Built 1951, constructor's number 12892, general purpose helicopter, powered by one Alvis Leonides 73 piston engine of 520 horse power.

This was the seventh production Sycamore and first flew at Filton on April 30th 1951 as G-ALSX. It was used by Bristols as a demonstrator and appeared at the Paris Air Salon at Le Bourget in June 1951 and the Society of British Aircraft Constructor's Display at Farnborough that September. From December it was used for aircraft carrier landing trials on H.M.S. 'Triumph'. From May 1952 it set off on a European demonstration tour and was to play a significant part in the relief work for the victims of the Dutch flood disaster in February 1953 - this was to open many peoples eyes to the value of the helicopter. In June 1953 G-ALSX was back at that year's Air Salon at Paris. From 1954 to 1957 G-ALSX settled back into the role of communications 'hack' for Bristol. In 1957 it was given a chance to renew its taste for the salt air with more carrier trials, this time on the Canadian carrier H.M.C.S. 'Magnificent'. In January 1958 G-ALSX was sold to the Williamson Diamond concern and the Tanganyikan registration VR-TBS was allocated to it but not used. In December 1959 it was to return to the makers, now under the aegis of the British Aircraft Corporation, this was short lived however as the Corporation sold off its helicopter interests, including G-ALSX, to Westland Helicopters in February 1960. Westland were to use the craft for communications and rotor blade experimental work until 1965. In May 1966 G-ALSX was registered as G-48/1. It was acquired by Mr. Elfan ap Rees in May 1969 and was loaned to Skyfame from July. G-48/1 is now part of the British Rotorcraft Museum's collection, but is to remain on loan.to Skyfame at least initially.

Chilton DW.1 G-AFGI

Built 1938, constructor's number DW1/3, single seat sporting aircraft, powered by one Lycoming O-145-A2 flat four piston engine of 55 horse power.

The third Chilton racer, G-AFGI first flew under the power of a Carden-Ford in-line of 32 horse power in 1938. Throughout its life it has been the subject of many modifications to enable it to perform better in races. It gained an enclosed bubble canopy with J. S. Sproule in 1950 and was extensively modified the following year by Hugh Kendall. In 1958 it was withdrawn from use and was dismantled at Scone, only to be spotted by G. Baker who bought it and by 1960 had it flying again. In 1966 it passed to J. J. H. Dickson of Marlborough who installed the Lycoming mentioned above. Mr. Dickson loaned the aircraft to Skyfame in 1968.

Cierva C.30A G-ACUU

Built 1934, constructor's number 726, single seat autogiro, powered by one Armstrong Siddeley Genet Major radial piston engine of 140 horse power.

Built by Avro as a type 671 under licence from Cierva G-ACUU was used before the War by Air Service Training at Hamble, but its certificate of Airworthiness was not renewed in May 1938 and it was put into store until impressed on September 2nd 1942. It was overhauled and issued to No.1448 Flight at Halton as HM580. Here it served on calibration and other duties alongside purpose-built Avro Rota Is. By June 1943 the Flight had been renumbered No.529 Squadron and HM580 was still with the unit. On October 18th 1943 its career nearly ended when it was declared Category E (write off) after an accident, indeed it was struck off charge that day. Cunliffe-Owen rebuilt HM580 at Eastleigh by January 28th 1944. On November 28th 1945 it was taken to No.5 Maintenance Unit at Kemble for disposal. In November 1946 it was registered as G-AIXE to the Cierva Autogiro Company and was overhauled ready for sale on the civilian market. During overhaul it was discovered, deep down under the camouflage, that this aircraft was really G-ACUU, for the original paperwork had been lost for the craft. The error revealed , G-AIXE was cancelled and G-ACUU restored on December 3rd 1946. In April 1950 it was sold to G. S. Baker, who flew it from Elmdon until 1955, when it went into store there. Mr. Baker loaned G-ACUU to the Museum in the mid '60s.

De Havilland DH.98 Mosquito TT. Mk.35 TA719 'SY-G'

Built 1945, high speed target tug, powered by two Rolls-Royce Merlin 114A piston engines of 1,710 horse power each.

Built in early 1945 by De Havilland's at Hatfield as a B. Mk.35 bomber variant, but was soon converted for target towing duties as a TT. Mk.35. It served with No.3 Civilian Anti-Aircraft Co-operation Unit at Exeter, coded '56' from August 1950. It was then used by Mirsch Films for the '633 Squadron' for which purpose it was registered as G-ASKC. In the film it was marked as 'HJ898' and coded 'HT-G'. After the film it became the property of the Museum, but was badly damaged in a belly landing during single engined flying on July 27th 1964. It has since been restored to static display standards and is coded 'SY-G' to commemorate the aircraft flown by Air Vice Marshal Bob Bateson.

De Havilland DH.112 Sea Venom FAW. Mk.22 WM571 '742/VL' see photo 91

Built 1954, two seat all weather naval fighter, powered by one De Havilland Ghost 105 turbojet of 4,850 pounds static thrust.

First flown at Christchurch in May 1954 as an FAW. Mk.21, WM571 was delivered to the Royal Navy Handling Unit at Boscombe Down on March 23rd 1955. On May 6th it joined No.891

Squadron. It was in store at Stretton from July 1957 until January 1958 and stored at Abbotsinch from July 1958 until January 1959. On November 1st 1959 it was returned to the makers for conversion to FAW. Mk.22 standard. It was flown to Culdrose in July 1963 where it joined No.831B Squadron and later became coded '384' (by September), later being '381'. In January 1964 it moved with the unit (now having dropped the 'B' suffix) to Watton having had all its electronic countermeasures gear installed. It was with Shorts at Sydenham from July 1965 until January 1966, returning to No.831 Squadron. In September 1967 it joined the Air Director School at Yeovilton and was coded '742/VL'. It was flown into Staverton on April 30th 1969 for the Museum and officially transferred to them on May 5th. In 1975 its restoration was taken on by Gordon Preston and Alan McIlray who plan to move the machine to nearby RAF Innsworth to work on it.

De Havilland DH.115 Vampire T. Mk.11 WZ515 '40'

Built 1953, constructor's number 15102, advanced trainer, powered by one De Havilland Goblin 35 turbojet of 3,500 pounds static thrust.

WZ515 was built at Hawarden and was officially handed over to the Royal Air Force on March 27th. It was delivered to No.22 Maintenance Unit at Silloth on April 1st. Operational service came first with No.16 Squadron as a continuation trainer, based at Celle in West Germany. It was flown to No.27 MU at Shawbury on June 6th 1956 for reallocation. It next joined No.253 Squadron at Waterbeach, being transferred to No.56 Squadron, also at Waterbeach, from September 23rd 1957. On April 14th 1959 it was delivered to No.8 Flying Training School at Swinderby, briefly being loaned to No.4 FTS as '40'. It was retired and delivered to No.19 MU at St. Athan on August 8th 1963 for disposal. It was bought by Hawker Siddeley Aviation on December 4th 1968 and removed to Hawarden for storage, later moving, in 1971/2 to Woodford. It was offered by HSA to the British Aircraft Preservation Council and acquired by Skyfame. It was delivered by road to Staverton on March 19th 1973.

Fairey Firefly I Z2033

Built 1944, constructor's number F.5607, target tug, powered by one Rolls-Royce Griffon XII piston engine of 1,900 horse power.

Z2033 was built at Heaton Chapel and test flown at Ringway before delivery to the Fleet Air Arm. It was disposed of in March 1949 to Svensk Flygstjanst AB of Sweden who were to use it for target towing work on contract to the Swedish Air Force. It was registered as SE-BRD and painted bright yellow. After service in Sweden it was donated to the Museum and flown in, in Swedish colours, on May 5th 1964 from Gothenburg. It was registered in the United Kingdom as G-ASTL and repainted as Z2033 and named 'Sir Richard Fairey'. It was maintained in airworthy condition, but at present is only capable of being taxied.

Hawker Tempest II LA607

Built 1943, prototype fighter, powered by one Bristol Centaurus V radial piston engine of 2,520 horse power.

LA607 was the second prototype Tempest, and flew with the radial Centaurus engine, the later production Tempest standardising on the Napier Sabre. It was used for engine and airframe development for all its life and was allocated to the College of Aeronautics at Cranfield in the late 1940s as an instructional airframe. It was presented to Skyfame in July 1966 by the College. It was repainted by the Museum in the colours of No.501 Squadron. In 1976 it was lent to No.41 Squadron at Coningsby and painted in the unit's colours, coded 'HF-T', to take part in its 60th Anniversary celebrations. It returned to Skyfame in late 1976.

Handley Page HP.67 Hastings C. Mk.1A TG528

Built 1947, long range transport, powered by four Bristol Hercules 216 radial piston engines of 1,800 horse power each.

TG528 was ready for collection by the Royal Air Force from Radlett on October 7th 1948. It was delivered to No.47 Squadron seven days later. On July 18th 1949 the unit moved to Topcliffe. TG528 suffered a Category 3 accident in early 1952 and a team from No.60 Maintenance Unit, Rufforth, arrived on January 16th to effect a repair. After much work, the team re-categorised the 'Haystack' as Category 4 (beyond repair on site). Handley Page were the best people to undertake the debending of the aircraft, but at the time could not accept the work, so TG528 was moved to No.5 MU at Kemble for storage from March 28th. Handley Page accepted it for work on May 23rd and finished on June 30th the following year. It was delivered to the pool used by Nos. 53 and 99 Squadrons, but TG528 was to be dogged by accidents, suffering another Category 3, being repaired by a team from Handley Page

84 *Superb air-to-air of Shuttleworth's Tomtit K1786, August 1971* *'Flight' L71-1493/7a*

85 *'Miss Kenya' - DH.51 G-EBIR - near Old Warden, August 1973* *'Flight' L73-1342/11*

86 *Comet K5084 (the famous G-ACSS) taxies at the RAF Pageant 1935* *'Flight' 11650 S*

87 *Halifax PN323 at Radlett in the late 1950's, note the mocked-up single fin.*

88 *The Gugnunc at Radlett the same time as the Halifax, making one of its very rare public airings!*

from January 30th 1954 until February 3rd. By October 15th TG528 had again been bent, this time a crew from No.49 MU Colerne administering to its needs, finishing on November 3rd 1954. Three days later it was back on strength and was soon to wear the last three letters of its call-sign ('MOJAT') on its fin 'JAT'. It then was damaged again, this time Category 4 and it was moved to No.49 MU for repair from May 14th 1955. No.49 MU realised that the work was beyond them and it was transferred to Handley Page on the 24th, until repairs were completed on March 26th 1956. It was now issued to No.242 Operational Conversion Unit on April 5th and was coded 'R'. The pattern of life was not to be changed with the change of unit though, and TG528 went Category 3 in a flying accident on January 9th 1957. No.60 MU were called in, starting work on the 15th and finishing on April 9th. It was back with No.242 OCU on the 11th and survived until January 23rd 1961 before its next Category 3. No.60 MU patiently moved in on February 27th, but had to hand over to a Handley Page crew on March 1st, who finished the job on July 10th. TG528 re-joined No.242 OCU on July 12th, and received the code 'J' some time afterwards. It was delivered to Handley Page, once again Category 3, on October 10th 1963 and was completed on June 30th 1964. On July 2nd 1964 it was issued to the pool of Nos 24 and 36 Squadrons, moving on to the charge of just No.24 Squadron on July 1st 1967. It was bought by John Fairey, Peter Swettenham, Bill Betteridge and the Skyfame Supporters Society in 1968 and flew into Staverton on January 24th 1968.

Mignet HM.14 Pou du Ciel BAPC.77

Built 1974, static replica, fitted with one converted Norton motorcycle engine.

This aircraft was built to original Henri Mignet plans by Michael Hayden and Ian Juggins for the Museum, starting in 1974, and finishing in 1975. It is fitted with a converted motorcycle engine. It has been allocated the British Aircraft Preservation Council indentification number 77.

Miles M.14A Magister G-AFBS

Built 1939, constructor's number 539, basic trainer, powered by one De Havilland Gipsy Major in-line piston engine of 130 horse power.

G-AFBS was used by its builder, Phillips and Powis Aircraft Limited for communications and training, based at Woodley. From the beginning of the War it was taken on by No.8 Elementary and Reserve Flying Training School, which was run by the firm, also at Woodley. It was impressed on September 17th 1940 as BB661. It was taken on charge by No.32 Maintenance Unit at St. Athan on November 20th 1941 and used for communications. It went back to Woodley on July 21st 1942, this time for service with No.10 Flying Instructors School. Here it was coded 'FDT-A' and served until May 28th 1946 when it was ferried to No.51 MU at Lichfield for disposal. It was sold in 1948 to British Overseas Airways Corporation, who registered it as G-AKKU in June. It was for the use of Airways Aero Associations (the flying club of BOAC) and the 'Maggie' was discovered to be G-AFBS on overhaul, G-AKKU was cancelled and G-AFBS restored. It was sold to the Denham Aero Club in October 1952. It continued to fly on a regular basis and was sold in 1962 to T. P. H. Jones of Blackbushe. It was secured for the Museum by Graham Johnson and it was restored to its early War-time hue of camouflage and yellow undersides, with civil registration.

Percival Proctor III LZ766

Built 1944, line number H.536, light communications aircraft, powered by one De Havilland Gipsy Queen in-line piston engine of 205 horse power.

This aircraft was built at Trafford Park, Manchester by F. Hills and Sons, and test flew at Barton as LZ766. In 1949 it was demilitarised and civilianised as G-ALCK. It was withdrawn from use at Woolsington in June 1963 and was taken over by an Air Training Corps unit in Tamworth, Birmingham. It was loaned to the Museum from this unit in 1965.

Saunders Roe SR.A.1 TG263

see photos 90 & 95

Built 1946, single seat flying boat fighter, powered by two Metropolitan Vickers Beryl turbojets of 3,500 pounds static thrust each.

The A.1 was designed in the later part of the Second World War to ultilise the new technology of jet propulsion and also as a means of over-coming the obvious vulnerability of runways. It was the first jet propelled flying boat in the world. Construction was started at the dispersed SARO plant in Beaumaris, in North Wales. It was moved to Cowes for finalassembly and was flown for the first time, as TG263, on July 16th 1947. The type was not put into production despite a very promising performance. SARO continued to fly the craft for a while as G-12-1, but eventually sold it to the College of Aeronautics at

Cranfield for use as a design example. One of its engines was then donated to Donald Campbell to power his ill-fated 'Bluebird' boat. It was donated to Skyfame in 1966.

Short SB.4 SHERPA G-36-1 see photo 92

Built 1953, constructor's number SH.1604, aerodynamic research aircraft, powered by two Turbomeca Palas turbojets of 350 pounds static thrust each.

The SHERPA (standing for Short and Harland Experimental and Research Prototype Aircraft) was developed from the SB.1 glider (constructor's number SH.1602) to carry on research into tail-less aircraft fitted with aero-isoclinic wings. It used the wings and tail section of the glider. The new aircraft, test registered as G-14-1, first flew at Aldergrove on October 4th 1953. In April 1957 it was taken on by the College of Aeronautics at Cranfield and re-registered G-36-1. It was used by the Bristol College of Advanced Technology as an instructional airframe from January 1964. In May 1966 it was moved to Staverton and is on display minus its wings.

Note : The Museum also has the following nose section on display :-

Handley Page HP.61 Halifax A.VII PN323 built by Fairey Aviation at Errwood Park, Stockport, in 1943. It was used, statically at Radlett post-War by Standard Telephones and Cables Limited for radio aerial experiments. During these it was fitted with a dummy single fin, in place of the normal twin unit. It was reduced to produce in the late 1960s and the nose section was presented to the Museum by Graham Trant. *(see photo 87)*

STRATHALLAN COLLECTION

The Strathallan Aircraft Collection was begun as recently as 1970 - almost by accident! Sir William Roberts only wanted to obtain a Hurricane when G-AWLW was advertised for sale in the magazine 'Flight', along with two two seater Spitfires (G-AVAV and G-AWGB) and a quantity of spares. This was in December 1969, and the aircraft had participated in the making of the film 'Battle of Britain'. As the vendor was unwilling to accept offers for individual aircraft, after much negotiation, the whole package was bought, and the Collection had started. In April 1970 these three aircraft were flown down to Shoreham for general overhaul. Some of the spare parts were also transported to Shoreham. Later, in the summer of 1970, one of the Spitfires (G-AWGB) was sold and shipped to Canada, and now flies under the appropriate registration CF-RAF.

In 1971 it was decided to base the aircraft at Strathallan airfield and the Spitfire was flown in in the December of that year - the same month as the AT-6 arrived by air from Cannes. The Hurricane was transported north by road in March 1972. By this time the Spitfire and the newly arrived Harvard IIB were the only two airworthy aircraft, although work was being carried out on the Hurricane, and the AT-6, both of which are now flying.

The story since these early beginnings has been one of incredible expansion in both aircraft and facilities at Strathallan. 1977 should see the opening of a huge new hangar, which will mean that the exhibits can all be housed indoors.

The Collection is kept in an airworthy condition as much as possible and it is planned to stage regular flying days in the long term. For the moments these are not possible as the needs of the many exhibits should be catered for first. Aircraft under restoration on the airfield are generally available for inspection, but some aircraft are stored away on the estate, or elsewhere on the airfield and are not open to view. The number of aircraft stored away from the airfield is always being reduced as more and more start their restoration programmes.

The Collection is open to the public from April 1st to October 31st from 1000 to 1700 hours. Admission currently stands at adults 50p and children under 12 free, car parking 10p.

Address : The General Manager, Strathallan Aircraft Collection, Strathallan Airfield, Auchterarder, Perthshire, Scotland, PH3 1LA.

Avro 652A Anson C. Mk.19 VM360

Built 1947, general purpose communications aircraft, powered by two Armstrong Siddeley Cheetah XVII radial piston engines of 420 horse power each.

VM360's service career was relatively short as Ansons go, from 1947 until retirement in 1957. It was sold to Hants and Sussex Aviation Limited and registered to them as G-APHV on September 5th 1958. By August 1960 it was with BKS Air Survey Limited and operated out of Rochford on surveys as far away as North Africa. It was to stay in air survey work for the rest of its commercial life, moving on to Survey Flights Limited at Newtownardsin June

1964 until becoming part of the Anson fleet of Kemps Aerial Surveys Limited, based at Thruxton in March 1969. In 1972 Kemps ceased trading and Sir William Roberts bought up their entire Anson fleet and G-APHV flew into Strathallan on February 5th 1973. It was repainted in its 1947 colours and is currently undergoing a full overhaul.

Avro 683 Lancaster X G-BCOH

see photo 97

Built 1945, long range heavy bomber, powered by four Packard Merlin 224s of 1,640 horse power each.

G-BCOH was born in 1945 at the hands of Victory Aircraft Limited at Malton, Canada as KB976. It was ferried to the United Kingdom and joined No.405 Squadron, based at Gransden Lodge, later at Linton-on-Ouse. In 1946 it was returned to Canada and was stored at Malton by Avro Canada Limited. In 1948 it was taken over by the Royal Canadian Air Force. It was refitted in 1953 for maritime reconnaissance status and served on with the RCAF in this role until it was struck off charge on April 9th 1964. KB976 was the last operational military Lancaster in Canada. It then took on the role of a fire-fighting water-bomber, and was registered as CF-TQC for the purpose. It was finally retired from this seasonal occupation in 1969. It was stored and bought by Sir William Roberts in 1974. On May 6th 1975 it was flown from Edmonton to Toronto after much work had been done to bring the Lancaster back to airworthiness. At Toronto its new identity, G-BCOH, was applied, as was the name 'Spirit of Caledonia'. The following day the ferry flight across the 'pond' started. The route was Toronto-Gander-Reykjavik-Abbotsinch, the latter being reached on May 20th. It was not until June 11th that G-BCOH touched down on Strathallan's grass. Work is now underway to bring it up to exhibition standards.

Avro 696 Shackleton T. Mk.4 VP293

Built 1950, maritime crew trainer, powered by four Rolls-Royce Griffon 67 piston engines of 2,450 horse power each.

This aircraft really came to fame with the Royal Aircraft Establishment at Farnborough where it was known with considerable affection as 'Zebedee' or even 'Zebers'. This came about for several reasons : it was part of the RAE's Weapons Flight and many of the little black boxes that it carried had to be harnessed inside the Shackleton's notoriously noisy and shakey fuselage on springs : few were the pilots who could settle the big tail-wheeled aircraft down without at least one bounce : and finally its pilots were all sworn fans of the BBC TV series 'The Magic Roundabout'. The character from which VP293 takes its name, Zebedee, is proudly portrayed on the aircraft's sides and the extreme rear to this day. VP293 started life as a standard MR. Mk.1 maritime reconnaisance bomber with No.236 Operational Conversion Unit, later joining No.224 Squadron based at Gibraltar. It ended this phase of its life at St. Eval, with No.42 and then No.206 Squadrons. It was then converted to a T. Mk.4 trainer and joined the Maritime Operational Training Unit until joining the Weapons Flight of the RAE in 1964. Retired from service in 1975 'Zebedee' flew into Strathallan on May 3rd 1976.

Bristol Bolingbroke IVT 9940

Built 1942, target tug, powered by two Bristol Mercury XV radial piston engines of 920 h.p.

This aircraft was built by Fairchild in Canada under licence from Bristol, the Bolingbroke being the name given to the Canadian version of the Blenheim. It was taken on charge by the Royal Canadian Air Force, serial 9940, on August 8th 1942, and served with No.5 Bombing and Gunnery School until it was involved in a Category B - beyond repair on site - accident on February 21st 1943. It was struck off charge with the RCAF on August 21st 1946. It was then stored with Wes Agnew until being bought by Sir William Roberts in 1972. It arrived at Strathallan, crated, on December 19th that year. Steady progress has been made on its restoration, but it will be a long time before it flies again.

De Havilland DH.80A Puss Moth VH-UQB

Built 1930, constructor's number 2051, light cabin monoplane, powered by one De Havilland Gipsy III in-line piston engine of 120 horse power.

This aircraft was built at Stag Lane in 1930 as G-ABDW. By October it had been sold to an Australian customer and it was flown out, becoming VH-UQB in May 1931. Its life out in Australia is not known in any detail. It is known to have been stored by Marshalls at Bankstown for some years towards the end of its life 'down under'. It was bought by the Collection and arrived at Strathallan in a crate in the early part of 1974. It has since been unpacked and awaits its turn for restoration.

De Havilland DH.82A Tiger Moth DF155

Built 1942, constructor's number 85904, basic trainer, powered by one De Havilland Gipsy Major in-line piston engine of 130 horse power.

DF155 was built by Morris Motors and entered service in 1942 with No.28 Elementary Flying Training School at Wolverhampton, now far from the Cowley works where it was built. In June 1947 the unit was renumbered No.25 Reserve Flying School. It was disposed of in 1953 and sold to the Association of British Aero Clubs, who had bought the machine on behalf of the Orkney Flying Club. It was registered as G-ANFV on December 1st of that year. By February 1962 it was owned by J. B. G. Campbell, and it carried the name 'Joes Burd'. On April 4th 1964 it crashed at Turriff and was cancelled from the register on August 24th. By November 1969 it was at Strathallan with Strathallan Air Services Limited and officially passed to the Collection on April 3rd 1970. It was restored to its original colours of bright yellow and has joined the fleet of flying aircraft in the Collection.

De Havilland DH.82A Tiger Moth G-AOEL

Built 1940, constructor's number 82537, basic trainer, powerplant as DF155 above.

Built in 1940 as N9510 this Tiger was first issued to No.24 Maintenance Unit at Ternhill on March 6th 1940. On May 7th of the following year it joined its first operational unit in the form of No.7 Elementary Flying Training School (later just No.7 RFS from 1947). On January 29th 1948 it went to No.1 RFS, followed by No.11 RFS from June 23rd 1950. On May 17th 1951 it had been temporarily retired to No.10 MU at Hullavington, before being issued for work with No.2 Grading Unit from September 2nd 1952. In March 1953 it was back in store, this time with No.33 MU at Lyneham, before being issued in July to No.7 Flying Training School at Cottesmore. On October 2nd 1953 it was retired to No.9 MU at Cosford, and disposed of, being officially recorded as sold to R. A. Peacock Limited on May 10th 1954. In September 1955 it was registered as G-AOEL and was to remain active until 1972. It had a string of owners, two of which spring to mind, Frank Hayter in 1962, and R. G. 'Dick' Emery in 1963. It was acquired by the Collection from Sywell and arrived on December 12th 1975 for rebuild to flying condition and possible export.

De Havilland DH.89 Dragon Rapide 4R-AAI

Built 1940, constructor's number 6736, feederliner/air taxi aircraft, powered by two De Havilland Gipsy Queen 3 piston engines of 200 horse power each.

This aircraft started life as a Dominie navigation trainer, built at Loughborough as NF865 by Brush Coachworks Limited. It was demobbed in May 1950 and registered as G-ALXT, but this registration was cancelled the following year when it was sold in Ceylon. It was registered in its new home as CY-AAI until Ceylon was re-organised in 1970 and it was re-registered 4R-AAI. It had been in continuous use as an air taxi and charter aircraft during that time. It was brought back to the United Kingdom in 1974 and was stored at Staverton for some time until bought by the Collection and flown to Strathallan on August 4th 1975.

De Havilland DH.94 Moth Minor G-AFPN

Built 1939, constructor's number 94044, two seat sporting monoplane, powered by one De Havilland Gipsy Minor in-line piston engine of 90 horse power.

This aircraft, built at Hatfield, was registered as G-AFPN on July 1st 1939. It barely had a pre-War flying career as a civilian, being impressed for War service on February 2nd 1940, with the serial X9297. It was with Nos.24 (Ternhill) and 20 (Aston Down) Maintenance Units respectively before being issued to No.7 Air Gunners School on April 14th 1942, being 'sub-let' to No.1 Air Armament School at Manby from March 20th until being disposed of at No.5 MU Kemble on May 31st 1945. It was restored as G-AFPN on February 12th 1946. It had several owners and was the subject of a major rebuild at Lasham in the mid 1960s. It was bought by the Collection and flown into Strathallan in May 1976.

De Havilland DH.98 Mosquito TT. Mk.35 RS712

Built 1946, target tug, powered by two Rolls-Royce Merlin 114A piston engines both of 1,710 horse power each.

RS712 was built as a bomber B. Mk.35 variant at Christchurch in late 1946 by Airspeed. It was taken to Brooklands Aviation at Sywell and converted to TT. Mk.35 standards. It first joined No.1 Civilian Anti-Aircraft Co-operation Unit, based at Hornchurch. It then moved to the Armament Practice Station at Schleswigland in northern West Germany. RS712 then entered a period of inactivity in store at No.27 Maintenance Unit at Shawbury until being delivered to No.3 CAACU at Exeter on April 29th 1958. Here it was coded '50'. After retirement it was sold to Mirsch Film and registered as G-ASKB for the film '633 Squadron' in which it played 'HJ682'. After the film it became the property of H. A. Mahaddie and was stored at Henlow and later at West Malling. It was restored and again prepared to become a movie star, this time in 'Mosquito Squadron', this time as 'RF580'. It was returned to West Malling after the filming and was bought by Sir William Roberts in September 1972. Much work was needed to restore it to flying condition, and it could not be ferried to Strathallan until November 8th 1975, still as 'RF580'. Once at Strathallan RS712 was given a full restoration job and emerged in a superb all silver scheme, representing the aircraft as it was on roll-out from the factory in late 1946.

De Havilland DH.106 Comet 2R XK655

Built 1953, constructor's number 06023, electronic counter-measures and reconnaissance aircraft, powered by four Rolls-Royce Avon 117 turbojets of 7,300 pounds static thrust.

This aircraft first flew as G-AMXA from Hatfield on August 27th 1953, and was delivered to British Overseas Airways Corporation at Heathrow on January 21st the following year. It was used for makers trials on radar and other installations and did not actually serve with BOAC. In 1963 work was completed thatturned theComet, now XK655, into an electronic counter-measures aircraft for use with No.51 Squadron at Wyton. It was replaced in 1972/3 by Nimrod R. Mk.1s and was retired, being initially stored out on the airfield at Wyton. It was acquired by the Collection for use as a centre point for the exhibits and certainly not to bring back to airworthy condition! It was flown into Strathallan's small (for a Comet) runway on August 21st 1974 in a most spectacular manner - the pilot succeeded in removing the starboard undercarriage from the rest of the aircraft on landing! Thankfully little damage was done to either flesh or aircraft and the Comet was soon erected and placed on display.

Fairey Battle I 1899

Built 1940, target tug, powered by one Rolls-Royce Merlin IIIpiston engine.

1899 was built as a standard Battle for the Royal Air Force by Austin Motors, serial R3950. It was transferred to the Royal Canadian Air Force and was shipped off to Canada, arriving there on July 1st 1941. It was converted to target tug status by No.8 Repair Depot at Winnipeg, serial 1899, it served throughout the war towing targets for No.2 Training Commande Group, RCAF. It was then stored and remained so until bought by Sir William Roberts and shipped to the United Kingdom. It arrived at Strathallan in a crate on November 5th 1972, and was put into storage on the estate. It was moved to the airfield very early in 1977 and restoration work is now underway on it.

Fairey Firefly AS. Mk.6 WD833

Built 1947, constructor's number F.8661, anti-submarine strike aircraft, powered by one Rolls-Royce Griffon 74 piston engine of 2,020 horse power.

WD833 was built to the order of the Fleet Air Arm, but was quickly transferred for service with the Royal Australian Navy, who used the same serial for it. Its service history is at present unknown. After disbanding from the RAN it was left to lie in increasing dereliction in southern Australia, until it was bought by the Collection and shipped back to the United Kingdom. It arrived in a crate at Strathallan on December 9th 1974. It has since been unpacked, but no work has been performed on it. Any restoration work on this aircraft will take a considerable length of time.

Fokker S.11 Instructor PH-ANK

Built 1952, constructor's number 6274 (but see below), basic trainer, powered by one Lycoming O-435-A flat four piston engine of 190 horse power.

PH-ANK was first taken on charge with the Dutch Air Force on September 26th 1952, and was issued to the Elementaire Vlieg Opleiding at Gilze-Rijen, with the serial E-31. It was later transferred on loan to No.9 Squadron of the Dutch Navy for the basic training of naval pilots, based at de Kooij, and on January 1st 1970 the allocation became permanent when the aircraft received the Navy serial 174. It was officially taken on charge with the Navy on the 26th. 174 remained at de Kooij with No.9 Squadron, fin coded 'K', until the type was withdrawn. No.9 Squadron disbanded in September 1973 and the S.11 went to the storage facility at Gilze-Rijen. It was later sold as PH-ANK and acquired by the Museum, arriving at Strathallan on January 7th 1977. The aircraft was built with the constructor's number quoted above, but is bound to have been converted to S.11 Series 1 standard with a new constructor's number in the 77.. series. Details welcomed.

General Aircraft GAL.42 Cygnet II G-AGBN

Built 1941, constructor's number 111, two seat cabin monoplane, powered by one Blackburn Cirrus Major II of 150 horse power.

The Cygnet was built at Hanworth and was impressed into War service almost immediately, on July 2nd 1941. It was delivered that day to No.23 Squadron as it had something very rare to offer in the training of crews for the new American aircraft that the unit was getting ready to operate. The rare commodity was a tricycle undercarriage to prepare tail-wheel trained pilots to fly the Douglas Boston bomber. Shortly after this it was allocated the serial ES915. ES915 was restored as G-AGBN in 1946 and went on to have a long and active civilian career, and is now the only one of its type left extant. For many years it was based at Biggin Hill and it was from there that it was acquired by the Collection in 1975, arriving at Strathallan on June 21st. In 1976 it was painted in its early Boston crew-trainer colours, still wearing G-AGBN.

Grumman TBM-3W Avenger 045

Built 1944, airborne early warning aircraft, powered by one Wright Double Cyclone 2600-20 radial piston engine of 1,750 horse power.

This bulky aircraft arrived at Dundee docks from Europort on May 16th 1976 and arrived at the airfield on the 18th. It is still in the dark sea grey and sky colours of its service with the Dutch Navy. It was first in use with the United States Navy, serial number 85650. It was delivered to the Dutch Navy under Military Defence Aid Procurement arrangements on September 29th 1953 and was given the serial P-102 and the code '16-102'. In 1959 it was re-serialled 045 in the re-arrangements of that year. It was struck off charge on November 14th 1960 and was sold to the Anthony Fokker School in The Hague, where it was used for technical instruction. It was bought from the School by C. Honcoop of Veen, and sold to Strathallan.

Hawker Hurricane IIB P3308 'UP-A'

Built 1942, constructor's number 42012, single seat fighter, powered by one Rolls-Royce Merlin 25 piston engine of 1,635 horse power.

This aircraft was built by the Canadian Car and Foundry Company in Montreal, Canada, for the Canadian Air Force, as 5588, with a Packard-built Merlin. Its history during the War and afterwards is not known until it was bought by Mr. R. E. Diemert of Manitoba in 1966 and registered as CF-SMI. Mr. Diement installed the Merlin 25 mentioned above and flew the aircraft for some time before selling it in the United Kingdom for use in the film 'Battle of Britain'. After the film it was sold, in July 1968, to N. A. W. Samuelson at Elstree, having been registered G-AWLW. It was the Samuelson concern who placed the advertisement in 'Flight' that the introductory narrative mentions. It became the property of Sir William Roberts in December 1969. Since then the aircraft has been completely overhauled and now flies as 'UP-A' P3308, representing the aircraft of Squadron Leader Archie McKellan of No.605 Squadron, who had 21 kills to his credit.

Lockheed Type 16 Hudson IV A16-199

Built 1942, constructor's number 6464, maritime reconnaissance bomber, powered by two Pratt and Whitney R-1830-S1C3G radial piston engines.

Lockheed's delivered this aircraft to the Royal Australian Air Force in 1942 and it flew with Numbers 2 and 13 Squadrons operationally with SEAC. After the War it was sold off and civilianised as VH-SMM and converted for use as a survey aircraft. It was later re-registered with Ad Astra Surveys as VH-AGJ and served with this firm for just under ten years. It was retired from use in the early 1970s and acquired by the Collection. It was flown the long way to the United Kingdom and touched down at Strathallan in May 1973. Since then it has been refurbished to static condition as A16-199, its original RAAF serial, and painted in representative SEAC roundels and camouflage, circa 1945.

Miles M.14A Magister R1914

Built 1940, constructor's number 1900, basic trainer, powered by one De Havilland Gipsy Major in-line piston engine of 130 horse power.

Built by Phillips and Powis Limited in 1940 for the Royal Air Force as R1914, this 'Maggie' has yet to reveal what War service it put in. It was civilianised in 1946 and registered G-AHUJ, taking the civilian designation of the Magister, the Hawk Trainer III. After a string of owners it was withdrawn from use in the mid-1960s and was dismantled at the gliding site of Aboyne. It was secured by the Collection in 1972, but it was not until mid

1975 that space was avialable and it was moved to Strathallan. Work started almost immediately on its restoration, as it was in very good condition considering the long time it had been in store. On June 26th 1976 the Magister flew again in its original hue.

Miles M.17 Monarch G-AFLW

Built 1938, constructor's number 792, cabin touring monoplane, powered by one De Havilland Gipsy Major I in-line piston engine of 130 horse power.

G-AFLW was registered to its builders, Phillips and Powis, at Woodley. It managed to escape impressment during World War Two for one reason or another. In March 1939 it was used by Rolls-Royce Hucknall and continued in the role of communications 'hack' for that concern for the rest of the conflict. Perhaps the Impressment Board felt that it was doing a vital enough job, or that they could not prise it out of the hands of the Rolls-Royce pilots! Post-War this beautiful cabin aircraft had several owners, notable being Derby Airways, based at Burnaston, from May 1958. In July 1968 it became the property of Rex Coates who set about a restoration to immaculate condition. Rex flew the machine regularly until sale to J. E. Randall in 1972. In 1976 it was acquired by the Collection, to join an impressive line-up of Miles types, and flew into Strathallan in June 1976.

Miles M.18 Series 2 G-AHKY

Built 1939, constructor's number 4426, basic trainer, powered by one Blackburn Cirrus Major III in-line piston engine of 150 horse power.

G-AHKY is the only surviving M.18. It was built at Woodley and flew for the first time in November 1939, under test conditions as U-9. It was then used by the Air Ministry for acceptance trials and was given the test registration U-0224. The trials came to nought and the M.18 was not put into production. U-0224 returned to Miles and was used by the firm for communications throughout the War, at first as U-0224, and then as HM545. In 1946 it was civilianised as G-AHKY. In 1961 it won the Kings Cup Race at an average speed of 142 miles per hour. This aircraft then entered a period of store, at Blackbushe, until, like the Monarch above, it came into the healing hands of Rex Coates. G-AHKY was moved to Heathrow for an extensive rebuild, flying again in 1973. In late 1975 it was bought by Sir William Roberts and flown up to Strathallan in mid 1976.

Miles M.38 Messenger 2A G-AJOC

Built 1947, constructor's number 6370, light cabin monoplane, powered by one Blackburn Cirrus Major 3 in-line piston engine of 155 horse power.

G-AJOC was first registered to C. E. Hickman of Wolverhampton. After a succession of owners it moved north to Scotland in 1969 in an unairworthy condition, and was put into store at Dunottar. It is thought that restoration was planned for this aircraft, but that it was never to come to fruition. By about 1972 this aircraft had been acquired by the Collection, but did not join the other items at Strathallan until 1973. It awaits restoration.

North American Harvard IIB FT391

Built 1943, constructor's number 14A-1431, advanced trainer, powered by one Pratt and Whitney R-1340-49 radial piston engine of 600 horse power.

Built by the Noorduyn Company in Canada, FT391 served with the Empire Training Scheme in Canada, being bought with Lend-Lease arrangements and allocated the United States Army Air Corps serial number 43-13132 for administration purposes. It was sold to the Dutch Air Force in the 1950s as B-97 and was later civilianised as PH-HON. It was sold in the United Kingdom and registered as G-AZBN in mid 1971, arriving initially at Shoreham. It was flown into Strathallan on April 2nd 1972. It now flies in typical wartime camouflage as FT391.

North American AT-6D Texan G-AZJD

Built 1943, advanced trainer, powerplant as FT391 above.

Built by North American for the United States Army Air Force, serial number 41-33931, with no traceable constructor's number. It was shipped to South Africa shortly after its commission, and joined the South African Air Force with the serial EX958. It was used under the Empire Air Training Scheme. Post-War it was sold to the Belgian Air Force as H-9. In the 1950s it joined the French civil register as F-BJBF and served for most of its life in France with the Air France School. It was bought by the Collection and flown to Shoreham in December 1971. It was registered G-AZJD and its overhauled commenced at Shoreham. It flew into Strathallan on July 17th 1972. It flies in the colours of a typical Texan trainer with the USAAF. (The Texan was the American name for the Harvard.)

Percival P.40 Prentice T. Mk.1 VS356

Built 1949, construction number 5830/3, basic trainer, powered by one De Havilland Gipsy Queen 30-2 in-line piston engine of 250 horse power.

VS356 was built by Blackburns and was first issued to No.22 Flying Training School on May 12th 1949, followed by No.2 FTS from March 19th 1951, and finally the Central Flying School from September 1st 1951. It was one of the hundreds of Prentices bought up by Aviation Traders Limited for possible conversion to civil status. VS356 was one of the few lucky ones, most ending their days on the scrap-heap. Registered as G-AOLU it was first flown after conversion on August 26th 1959 and was based at Biggin Hill until July 1967 when it was sold in Eire and registered as EI-ASP for use with Joyce Aviation. It was returned to the United Kingdom in October 1972 and spent some time in open storage at Baginton before finding a buyer and being restored as G-AOLU. It again took up residency at Biggin Hill before being acquired by the Collection. It was flown up to Strathallan on September 30th 1975. By August 1976 it had been resprayed in the silver and yellow scheme that it wore in its days with the RAF.

Percival P.56 Provost T. Mk.1 WV493

Built 1953, constructor's number P.56/56, advanced trainer, powered by one Alvis Leonides 126 radial piston engine of 550 horse power.

Built at Luton, WV493 was ready for collection by the Royal Air Force on October 28th 1953. It was to serve the RAF operationally only with one unit, No.6 Flying Training School to whom it had been issued two days previously. It was coded 'P-A' with this unit. It was retired to No.27 Maintenance Unit at Shawbury on June 20th 1957 and was declared a non-effective airframe on June 28th 1960. It was moved to No.1 School of Technical Training at Halton on November 16th and was given the instructional airframe number 7696M and the Halton code '29'. It was bought by the Collection in 1976 and allocated G-BDYG, flying into Strathallan, still carrying its 'M' number, on June 23rd 1976.

Reid and Sigrist RS.4 Desford VZ728 see photos 93 & 96

Built 1945, constructor's number 3, aerial survey aircraft, powered by two De Havilland Gipsy Major 1C-2 in-line piston engines of 145 horse power each.

This unique aircraft was built at Desford by the company in 1945, it was intended as a small twin engined conversion trainer, and was then powered by two Gipsy Major Is of 130 horse power. It was not adopted by the Royal Air Force and remained with the makers, as G-AGOS. In May 1949 it was modified for prone-pilot experiments. For this purpose the nose was lengthened and given a glazed forward end giving the second pilot, lying on his stomach, with special controls, a superb field of view. It was redesignated the RS.4 Bobsleigh and given the serial VZ728. It first flew in this form on June 13th 1951. It was used by the Institute of Aviation Medicine at the Royal Aircraft Establishment at Farnborough. After the tests the aircraft returned to being G-AGOS and was bought by Air Courier Limited. The firm overhauled it at Croydon and fitted the new engines noted in the heading and called the aircraft the Desford Trainer. In March 1958 it was bought by Film Aviation Services Limited and operated by them until September. With its nose clear of the second set of instruments and controls the aircraft made an ideal camera-ship. In 1963 it was taken on by the fleet of Kemps Aerial Surveys and used almost as the private mount of the owner of that company. It was sold off in 1972 and the new owner painted it in military guise with the serial VZ728 but a totally inaccurate War-time camouflage scheme. The Desford was bought by Sir William Roberts in 1975 and arrived at Strathallan in the middle of the year. It is hoped that it will be returned to more accurate colours.

Supermarine Spitfire XIV NH904

Built 1944, single seat fighter, powered by one Rolls-Royce Griffon 65 piston engine.

This Spitfire moved to Strathallan in the early weeks of 1977 from Sir William Roberts' farm at Flimwell, in Sussex, where it had been in store. Its restoration should take a long time to complete. This aircraft was bought by Sir William Roberts from Hamish Mahaddie and has been stored at Flimwell ever since purchase.
Details of this aircraft welcomed.

Westland Lysander III G-AZWT

Built 1942, army co-operation aircraft, powered by one Bristol Mercury radial piston engine of 890 horse power.

This aircraft was built in Canada by the National Steel and Car Corporation at Hamilton,

Ontario. It entered service with the Royal Canadian Air Force with the serial 2355. It was retired and was stored for many years in the open along with several others of the type by Wes Agnew. It was bought by Sir William Roberts in late 1971 and along with many spares it was crated up and shipped to the United Kingdom. It arrived at Strathallan on February 14th 1972, but was not uncrated until 1974. From then on a considerable amount of restoration work has been done on the airframe and it is now nearing completion and may well fly in late 1977. It was registered as G-AZWT in 1972.

Note : The following aircraft are in store on the estate or elsewhere on the airfield and are not available to public view at the time of writing :-

Avro 652A Nineteen Series 2 G-AGWE built for the RAF as Anson C. Mk.19 TX201, served with Kemps Aerial Surveys, flown into Strathallan on January 16th 1973. In external store.
Avro 652A Nineteen Series 2 G-AHIC built 1946, constructor's number 1318, served with Kemps Aerial Surveys, flown into Strathallan on January 16th 1973. In external store.
Avro 652A Nineteen Series 2 G-AHKX built 1946, constructor's number 1333, served with Kemps Aerial Surveys, flown into Strathallan on January 22nd 1973. In external store.
Avro 652A Nineteen Series 2 G-AWRS built for the RAF as Anson C. Mk.19 TX213, served with Kemps Aerial Surveys, flown into Strathallan on February 5th 1973. In external store.
Avro 652A Nineteen Series 2 G-AYWA built as OO-CFA, later OO-DFA, and OO-VIT of Belfotop. Imported by Kemps but not converted by 1972. In store dismantled on the estate.
Bristol B.171 Sycamore HR. 14 WA576 bought from No.1 School of Technical Training at Halton in 1974, where it was 7900M. In store on the airfield.
De Havilland DH.82A Tiger Moth G-ANTS built for the RAF as N6532, c/no. 3845. Not converted for civil use, to Strathallan by road on March 2nd 1972 from Mintlaw.
De Havilland DH.82A Tiger Moth G-APGL built for the RAF as NM140, history as for G-ANTS.
De Havilland DH.84 Dragon VH-SNB built for the RAAF as A34-13 by De Havilland Aircraft Pty. at Sydney. Civilianised as VH-ASK, later VH-SNB. Imported in mid 1974, still crated.
Short S.16 Scion II VH-UUP built as a seaplane G-ACUX, c/no. S.776, in 1934. To New Guinea in February 1938 as VH-UUP. Withdrawn from use in 1947 and preserved at Bankstown. Arrived at Strathallan in mid 1974. Stored on the estate.
Supermarine Spitfire Tr.IX ML407 Supplied to the Irish Air Corps as 162 in 1951, withdrawn from use in 1960. Used for spares in the 'Battle of Britain' film. Bought from Elstree in December 1969. Arrived at Strathallan on March 2nd 1972. In store on the estate.
Supermarine Spitfire Tr.IX PV202 History as for ML407, Irish Air Corps 161.

TORBAY AIRCRAFT MUSEUM

The Aircraft Museum at Torbay is unique amongst collections in the United Kingdom in its fervant emphasis on the small but fascinating memorabilia of aviation. Not that there is any shortage of aircraft - twenty or more are permanently on display - but the Directors of this self-supporting Museum are constantly aware of their obligation to visitors whose interests in aeronautical matters is strictly limited. With this in mind they have set about producing a display in their extensive indoor galleries which aims to stimulate the interest and attention of even the most reluctant visitor.

Although the Museum's 32-page souvenir guide book is a 'must' for the background details of the aircraft themselves, the dozens of items in the galleries are fully annotated, and in each case the wording of the captions is heavily slanted towards human interest as opposed to technical data. This is a Museum that believes that since humans created aeroplanes, those humans with their creativity, aspirations, disappointments, and the other dramatic blends of triumph and tragedy offer as much interest as the aeroplanes themselves.

For the visitor to South Devon, Torbay Aircraft Museum offers an outing which is 75% under cover if it is raining, and is sited in an area of outstanding natural beauty when the sun shines and the limitless views over Torbay to the east and Dartmoor to the west are revealed in all their glory. Admission is 50p for adults and 25p for children and toddlers under school age are admitted free. There is a licensed cafeteria during the peak season, the Museum shop has an extensive range of goods at all prices and car parking is entirely free. Opening time is 1000 hours daily throughout the year (except Christmas Day and Boxing Day) and closing time varies from 1800 (summer) to 1600 (winter). By car take the A385 Paignton/Totnes road, turning right at the Museum signpost at Collaton St. Mary, or if arriving by train both Paignton and Totnes stations are equidistant. A taxi fare from either station should cost under £1. During the peak summer season Devon General buses run a direct service to the Museum from Paignton Bus Station.

Address : Torbay Aircraft Museum, Higher Blagdon, near Paignton, Devon, TQ3 3TG
Telephone : 0808 553540 (24-hour recording service)

Avro 594 Avian III G-EBZM

see photo 94

Built 1928, constructor's number 160, two seat tourer, powered by one ADC Cirrus of 80 h.p.

G-EBZM was built at Woodford and gained its first Certificate of Airworthiness on August 3rd of that year. Apart from being built in the north-west, it has maintained its links with the area ever since. It first saw service with Merseyside Aero and Sports Limited, based at Hooton Park from February 1929. The majority of its active life was as a joy-riding mount of Giro Aviation, who flew from Southport Beach. G-EBZM served in this capacity from 1932 until retirement in 1938. It was stored in a dismantled state through the War and beyond in Giro's premises at Hesketh Park. In 1960 it was rescued by the Merseyside Group of Aviation Enthusiasts (previous name for the publishers of this book) and stored, firstly in Huyton, then at Speke, and then at Lymm Grammar School. This activity gave rise to the Northern Aircraft Preservation Society, now The Aeroplane Collection, and G-EBZM became one of their first aircraft. It was actively restored in the Manchester area, using parts from another Avian (G-ABEE) found by the group in November 1962. After much work it was suitable for display, and after some brief appearances at local air displays around Manchester it was loaned to Torbay for exhibition.

Bristol 171 Sycamore HR. Mk.14 XG544

Built 1956, constructor's number 13384, light helicopter, powered by one Alvis Leonides 73.

XG544 came off the line at Old Mixon in 1956 and was issued to No.275 Squadron at Leconfield, for search and rescue duties. No.275 Squadron was re-numbered No.228 Squadron on September 1st 1959. A year later XG544 was placed in store at No.23 Maintenance Unit at Aldergrove, as Whirlwind HAR. Mk.10s had replaced the type with the Squadron. In 1964 XG544 was delivered to the Metropolitan Communications Squadron at Northolt, for a new role as a VIP transport. In 1972 the MCS was re-numbered as No.32 Squadron. In that year XG544 was again replaced by the Whirlwind HAR. Mk.10 helicopter and it was struck off charge on August 10th. XG544 had the distinction of being the last Sycamore to fly in RAF service and it made one last flight from Northolt to Upavon and then on to Higher Blagdon to take its place in the Museum, the day after it was taken off RAF charge.

De Havilland DH.115 Vampire T. Mk.11 XE995

Built 1955, constructor's number 15675, advanced trainer, powered by one De Havilland Goblin 35 turbojet of 3,500 pounds static thrust.

XE995 was officially handed over to the Royal Air Force on June 30th 1955, at Hatfield. It was issued straight away to No.12 Maintenance Unit at Kirkbride, prior to moving into active use. On September 26th 1955 it was issued to No.32 Squadron (who were later to operate XG544 above) at Habbaniyah, Iraq, and later based at Malta. Again in store from January 16th 1957, this time at No.19 MU at St. Athan, and from April 23rd 1959 at No.5 MU, Kemble. The following month, the 27th, XE995 again started active service, with No.5 Flying Training School at Oakington. It served within the School with No.3 Squadron and gained the code number '63'. It continued with the flying training role with its next unit, No.8 FTS at Swinderby, from October 30th 1962, this time coded '53'. On March 11th 1964 it ended its flying days when it returned to No.19 MU. In 1968 Hawker Siddeley Aviation bought back the machine and stored it first at Hawarden, and then at Woodford, in the hope of a possible export sale. It was acquired by the Museum in 1973 and travelled by road to Higher Blagdon on March 23rd.

De Havilland DH.104 Dove Series 6 G-ALFT

Built 1948, constructor's number 04233, light transport, powered by two De Havilland Gipsy Queen 70 in-line piston engines of 340 horse power each.

G-ALFT served all its life with the various Government Departments responsible for the checking of aviation navigation aids, runway lights etc. all over the United Kingdom. It started life as a Series 2, but was progressively modified to Series 2B and later to a Series 6 aircraft. It was handed over to the Ministry of Aviation on January 18th 1949. Its long career saw it registered to the following masters :- the Minister of Transport and Civil Aviation from December 1953, the Board of Trade from November 1956, the Department of Trade and Industry from January 1971, and, finally, the Civil Aviation Authority from April 1972. It that year it was retired and moved to Higher Blagdon by road.

89 *Anson G-AMDA at Burnaston in 1958* — *P. H. Butler*

90 *SR.A.1 G-12-1 at Cranfield in 1966* — *P. H. Butler*

91 *Sea Venom WM571 at Yeovilton 9/9/67* — *Dennis Price*

92 *SHERPA G-14-1 at Farnborough 11/9/54* — *R. A. Scholefield*

93 *Desford G-AGOS on trestles at Speke*

94 *Avian G-EBZM at Hesketh Park, Southport* — *NAPS*

95 *Skyfame's SARO SR.A.1 TG263 gets airborne, Summer 1947* *'Flight' 20487 S*

96 *Desford G-AGOS in original configuration, over Leicester, 1948* *'Flight' C102/11*

97 *Strathallan's Lancaster at St. Albert, Canada, being restored* *Dick Richardson*

De Havilland DH.112 Sea Venom FAW. Mk.22 XG629

Built 1956, two seat all weather naval fighter, powered by one De Havilland Ghost 105 turbojet of 4,850 pounds static thrust.

This aircraft was taken on charge by the Fleet Air Arm at Stretton on March 27th 1956 and remained there until at least July. It was issued out for Squadron use at Yeovilton in January 1958 and toured the Mediterranean Sea from February 9th 1958, embarked on an unknown carrier with an unknown unit! (Details welcomed) It was involved in a touch down accident on land on July 25th 1958. It was issued to No.893 Squadron shortly afterwards, being given the side code '459'. It moved to Sydenham in January 1959 to be converted from an FAW. Mk.21 into the FAW. MK.22 variant. It was back in service with the Squadron, and one H.M.S. 'Victorious' by July. It served on with No.893 until joining No.831 Squadron in July 1964. This was after a period of modifications with Shorts at Sydenham and also at Lee-on-Solent. It served with No.831 Squadron, as '380', from July 1964 until January 1966. It was stored at Sydenham from July 1968 until February 1970. It then joined the Air Director School at Yeovilton, serving only until July 1970. It was then issued to the fire section at Culdrose for what should have been a hot death. It was bought by the Museum from Culdrose and arrived at Higher Blagdon in May 1971.

De Havilland Canada DHC-1 Chipmunk T. Mk.10 WB758 'P'

Built 1950, basic trainer, powered by one De Havilland Gipsy Major 10 of 145 horse power.

The 91st Chipmunk to be built at Hawarden, WB758 was completed in 1950 and delivered in November of that year to No.14 Reserve Flying School at Hamble. In July 1953 it moved for a brief period of storage to No.20 Maintenance Unit at Aston Down, before moving to the Transport Command Communications Flight, based at Andover, on October 19th 1953. In 1961 it was used for a while by Oxford University Air Squadron, Abingdon, before being struck off charge with the Royal Air Force in that year. It was issued as an instructional airframe to No.2030 Squadron, Air Training Corps, at Sheldon, Birmingham. At this time it was allocated the maintenance airframe serial 7729M. In 1970 it was used by No.71 MU at Bicester, as a demonstration and display airframe at various exhibitions across the country. It was bought by the Museum in 1972, and arrived at Higher Blagdon by road on October 21st that year.

Focke Achgelis Fa 330A-1 Bachstelze 100545

1944, single seat observation rotor-kite.

This example of an Fa 330 is on loan to the Museum from Mr. S. S. Miles and is displayed in the main building. It was acquired by Mr. Miles in the Bedford area. It is one of several brought over to the United Kingdom in 1945 and tested at either Farnborough or Beaulieu. It is almost certainly an Fa 330A-1.

Gloster Meteor T. Mk.7 (hybrid) WF877

Built 1951, advanced trainer, powered by two Rolls-Royce Derwent 8 turbojets of 3,600 pounds static thrust each.

WF877 was issued from its birthplace, Hucclecote, to the Royal Air Force on July 18th 1951. It first saw service with No.11 Squadron, based at Wunsdorf, West Germany, where it was used for continuation training alongside the unit's NF.Mk.11 fighter versions. In November 1952 it moved on to No.96 Squadron at Ahlhorn for similar duties. The following November it was flown to No.12 Maintenance Unit, Kirkbride, and its active service ended. In April 1954 it was delivered to Martin Baker Limited at Chalgrove, where it was used in development work for the company's famous ejector seats. It continued with ejector seat work with its next operator, Folland Aircraft at Chilbolton, where it was used to perfect the seat for the Gnat fighter/trainer. In June 1966 it moved to Tarrant Rushton airfield, the home of Flight Refuelling Limited, and WF877's flying days ended. Here it was used as a ground test bench for overhauled Derwent engines. When it was retired from these duties it was acquired by the Museum, and arrived by road on February 1st 1974.

Hafner R-II Revoplane BAPC.10

Built 1933, experimental helicopter, powered by one Salmson radial piston engine of 45 h.p.

A pioneering British helicopter designed by Austrians Raoul Hafner and Bruno Nagler, and test flown in 1933. Mr. Hafner went on to become the chief designer for the Sycamore helicopter (see above). The Revoplane was never registered. It was stored at Yeovil until 1966when it was donated to the Shuttleworth Trust. It was loaned to the Museum and arrived by road on August 11th 1972. It has been allocated the British Aircraft Preservation

Council identification number 10.

Hawker Sea Hawk FB.3 WM961

Built 1954, naval fighter bomber, powered by one Rolls-Royce Nene 103 turbojet of 5,400 pounds static thrust.

Built at Baginton by Armstrong Whitworth Aircraft, WM961 was first delivered to Abbotsinch on June 29th 1954, joining No.898 Squadron in September. From July 1955 it was with No.806 Squadron, coded '165/C' for a brief period before joining No.811 Squadron at Lossiemouth. With the unit it was to tour the Mediterranean on H.M.S. 'Centaur' and the Far East. In November 1956 it was converted from FB. Mk.3 status to FB. Mk.5 equipment level. It joined No.802 Squadron, as '135/O', in September 1957 and was again in the 'Med'. It was delivered to the Fleet Requirements Unit at Hurn on January 25th 1960 and stayed there until going to Arbroath for life as an instructional airframe, number A2517, in June 1961. From Arbroath it moved to Culdrose and took on the role of a deck handler trainer with the School of Aircraft Handling, being numbered SAH-6. It was acquired by the Museum from Culdrose and arrived in May 1971.

Miles M.38 Messenger 2A 'RG333'

Built 1947, constructor's number 6707, light cabin monoplane, powered by one Blackburn Cirrus Major III in-line piston engine of 155 horse power.

This aircraft was built by the Miles plant at Newtownards in Northern Ireland, in September 1947. It was allocated the civil registration G-AKEZ and it was delivered to its first owner, T. Shipside Limited, at Tollerton in 1947. It had several owners, but languished at Lulsgate after its last flight on September 13th 1957, having amassed some 1,410 hours in the air. It was moved to Torbay in May 1971. It has been repainted to represent RG333, the Messenger that served as Field Marshall Montgomery's private transport during World War Two.

Percival Proctor IV NP184

Built 1944, line number H.591, communications aircraft, powered by one D.H. Gipsy Queen 2.

Built in 1944 by F. Hills and Sons at Trafford Park, Manchester, for the Percival Company, this Proctor entered service with the Royal Air Force as NP184 on communication duties. In 1955 it was sold off as surplus and civilianised as G-ANYP with the Wiltshire School of Flying, at Thruxton. Its flying days finished in 1966 and it was used by the Brooklands Technical College as an instructional airframe at Weybridge. It was acquired by the Museum, and after being refurbished by the Apprentices of the Royal Aircraft Establishment at Farnborough, which involved returning it to its War-time colours as NP184, it arrived at Higher Blagdon in March 1972.

Percival P.56 Provost T. Mk.1 WV605

Built 1954, constructor's number P.56/132, advanced trainer, powered by one Alvis Leonides 126 radial piston engine of 550 horse power.

This is one of two Provosts on display at the Museum. WV605 was delivered from Luton to No.22 Flying Training School, Syerston, on March 21st 1954, being coded 'T-B' with the unit. Its stay here was short, moving on to No.3 FTS at Feltwell, on August 25th, this time taking the codes 'N-P'. Its final active charge was with No.6 FTS at Acklington, from June 15th 1957, coded 'O-V'. It was sold off to Macclesfield College of Further Education for use as an instructional airframe on October 31st 1966, after store with No. 27 Maintenance Unit at Shawbury. The College replaced the Provost with a Vampire T. Mk.11 and the Provost was donated to the Shuttleworth Trust. It was loaned to the Museum in 1971.

Percival P.56 Provost T. Mk.1 WV679 'O-J'

Built 1954, constructor's number P.56/199, other data as WV605.

WV679 followed WV605 (above) out of the Luton factory on September 14th 1954, but was to only serve actively with one unit. On that day it was issued to No.2 Flying Training School at Hullavington, later moving to Syerston. It was given the codes 'O-J'. It was declared a non-effective airframe in October 1959, moving to No.1 School of Technical Training at Halton, for mechanical instruction training. It was given the maintenance airframe serial number 7615M. It was bought by the Museum from No.1 SoTT and moved to Higher Blagdon on December 20th 1971.

Saunders Roe Skeeter AOP. Mk.12 XN351

Built 1960, constructor's number S2 7157, aerial observation post, powered by one De Havilland Gipsy Major 215 in-line piston engine of 215 horse power.

XN351 was awaiting collection by the Army Air Corps at Eastleigh on August 30th 1960 and moved to No.15 Maintenance Unit at Wroughton for storage prior to issue to a unit. On that date in 1961 it was issued to No.651 Squadron at Middle Wallop, moving in October 1963 to No.652 Squadron, based in West Germany. It was later moved to join No.3 Royal Tank Regiment, at Detmold, before returning to No.15 MU on June 20th 1967. It was declared a non-effective airframe on May 27th 1968 and was acquired by the Shuttleworth Trust on August 13th 1968, and it was kept at Old Warden for some time. It was loaned to the Museum in May 1971.

Westland WS-51 Dragonfly HR. Mk.5 WN499

Built 1953, constructor's number WA/H/76, general purpose helicopter, powered by one Alvis 521 radial piston engine of 500 horse power.

WN499 first flew, at Yeovil, on April 21st 1953, and was taken on charge by the Fleet Air Arm at Gosport on May 7th. By July it had joined the Ship's Flight of the aircraft carrier H.M.S. 'Illustrious', but was back at Gosport in July 1955. It had joined No.705 Squadron by at least July 1957, becoming '530'. On December 16th 1957 it was flown to Donibristle and converted from a HR. Mk.3 into a HR. Mk.5 Dragonfly. This work was completed on March 1st 1958. By January 1959 it was with the Station Flight at Eglinton and back with No.705 Squadron in July, becoming either '911/CU' or '913/CU'. It was struck off charge at Fleetlands on June 1st 1964 and was sold to the Anglo Diesel Company and moved to Blackbushe on April 28th 1965. At Blackbushe it became a familiar part of the landscape, along with several other Dragonflies, some of which are dotted about this book. It was acquired by the Museum in 1971.

Westland WS-55 Whirlwind HAR. Mk.3 XJ393

Built 1954, general purpose helicopter, powered by one Alvis Leonides Major 755 radial piston engine of 700 horse power.

XD363 (as this Whirlwind was originally numbered, but quickly altered to XJ393) started its hovering trials at Yeovil on September 30th 1954, being the HAR. Mk.3 prototype. It was not until June 27th 1955 that it first flew properly. It moved then to the Aeroplane and Armament Experimental Establishment at Boscombe Down for trials prior to the type's acceptance by the Fleet Air Arm. On March 1st 1956 it was delivered to the Royal Radar Establishment at Defford, during which time it did much development work. It moved with the RRE to Pershore on September 20th 1957. In 1960 it was returned to the Navy and reduced to instructional airframe status with Arbroath as maintenance airframe number A2538 and engine number E4566. It was struck off charge on January 28th 1966. XJ393 was transferred to the Air Engineering School at Lee-on-Solent in July 1970. It was acquired by the Museum from here and moved to Higher Blagdon in March 1972.

Note : The Museum also has the following aircraft, which is under restoration in the Museum workshop and is not available to public view :-

De Havilland DH.82A Tiger Moth G-ANSM built 1940, constructor's number 82909, built for the RAF as R5014, sold by the RAF in June 1956, becoming G-ANSM, flown into Higher Blagdon in 1971 and briefly displayed. Now under restoration to flying condition.

ULSTER FOLK & TRANSPORT MUSEUM

The Ulster Folk and Transport Museum is an off-shoot of the main Belfast Municipal Museum. Cultra Manor, near Holywood, to the north west of Belfast, has long been earmarked as the site of a Museum dedicated to recording and displaying the cultural and rural heritage of the Province. At the same time the chronic demands of space in the Belfast Transport Museum in Witham Street, Belfast, dictated that an annex for this display should be created. Thus the seemingly diverse title to this Museum. At present only the Ferguson Monoplane replica is on display at Cultra Manor, but it is hoped to expand the number of whole aircraft exhibits here in the future. The new Museum opened in 1976.

At the time of compilation, times of opening and admission fees are not known.

Address : The Ulster Folk and Transport Museum, Cultra Manor, Holywood, County Down.

Ferguson Monoplane replica

Built 1973, static replica, fitted with a JAP V-8 piston engine of 35 horse power.

On June 12th 1911 Harry Ferguson's fifth monoplane flew for the first time at Newtownards. These were the first aircraft to be made and flown in Ireland. The Ferguson Monoplane flew on till at least 1913. The Ferguson family still had the engine and the seat from this aircraft when they kindly donated them to the Irish Aviation Museum in December 1969. The donation of these parts set the Honorary Curator, Captain J. C. Kelly-Rogers, about the task of building a static replica of this highly significant aircraft. Work started in Dublin in the middle of 1972 and it was complete by early 1973. After several exhibitions in Eire, it was displayed as the centre piece of the Aeronautical Exhibition at Dublin Airport. In 1976 it was loaned to the Ulster Folk and Transport Museum and delivered to Cultra Manor by road on May 4th.

Note : The following aircraft are in store for the Museum :-

De Havilland DH.82A Tiger Moth G-AOUR built for the RAF as NL989, constructor's number unknown, crashed at Newtownards June 6th 1969, in store in Belfast.
Miles M.11A Whitney Straight G-AERV built 1937, constructor's number 307, impressed May 1941 as EM999, withdrawn from use in 1966, in store in Belfast.
Supermarine Spitfire LF.XVIe TE184 6850M once part of the RAF Finningley collection, coded 'ME-M', in store at Aldergrove.
Short SC.1 XG905 built Sydenham 1958, constructor's number SH.1815, first VTOL transition April 6th 1960, donated in May 1974, in store at Sydenham.

WALES AIRCRAFT MUSEUM

The Museum is situated on a two acre site alongside the main approach road to Glamorgan Rhoose Airport. The Museum was founded by the South Wales Aircraft Preservation Society, which in its turn was a founder member of the British Aircraft Preservation Council.

The Museum was planned in two stages :-

1) an open air display of metal skinned aircraft and historic vehicles.

2) an under cover site for the display of fabric skinned aircraft, aero engines, World War Two aero relics and general aero memorabilia.

Phase One was opened during September 1976, and the initial fee for entry was 15p. Opening times being limited to weekends. During Spring 1977 these will be revised.

Phase Two will house a quite extensive collection of aero engines including the following : a Rolls-Royce Merlin 30, a Rolls-Royce Merlin 70 for a Spitfire which crashed in Swansea in 1946, a Rolls-Royce Griffon 58 from a Shackleton, a Bristol Siddeley Viper 102 from a Jet Provost, a Bristol Hercules 264 Engine Change Unit from a Varsity, an Alvis Leonides Major test engine, an Alvis Leonides 126 Engine Change Unit from a Pembroke, a De Havilland Goblin from a Vampire, a Bristol Olympus 101 from a Vulcan, a Rolls-Royce Tyne from a Belfast, a Bristol Proteus 'exploded section' from a Britannia, a Rolls-Royce Dart from a Viscount and a Rolls-Royce Derwent from a Meteor. Along with these will be a collection of flying gear, World War Two relics, a collection of modern canopies, undercarriages and cockpit equipment.

Address : Wales Aircraft Museum, 3, Brynmead, Bryn, Llanellis, Dyfed SA14 8QQ.

Auster B5 AOP. Mk.9 WZ662

Built 1954, constructor's number AUS/B5/1 (unconfirmed), aerial observation post, powered by one Blackburn Bombadier 203 in-line piston engine of 173 horse power.

WZ662 was the prototype Auster AOP. Mk.9 and flew for the first time on March 19th 1954, from Rearsby. It was involved for the first part of its flying life in development flying, moving into active service in Janaury 1959 with No.2 Flight/No.656 Squadron. It was issued to No.19 Maintenance Unit at St. Athan on July 30th 1965, and was declared a non-effective airframe on November 8th 1968. It was sold to Mr. 'Dizzie' Addicott on October 15th 1969 and in turn sold to Toogoods Caravans on Bedminster Down near Bristol for display outside their garage. It was discovered in this condition by members of the South Wales Historic Aircraft Preservation Society (as SWAPS were then called) who arranged to swop this significant aircraft with another AOP. Mk.9 with Toogood's. This deal arranged, WZ662 was moved to St. Athan, where it was kept, and displayed at Battle of Britain days, whilst SWAPS members refurbished it to display standards. In February 1976 it was moved to Rhoose to take its place in the Museum.

De Havilland DH.112 Venom NF. Mk.3 WX788

Built 1954, two seat night figher, powered by one De Havilland Ghost turbojet.

This aircraft was ready for collection by the Royal Air Force on February 28th 1954. It was never to serve operationally however, devoting its flying life to manufacturer's trials. It was used by De Havilland's at Hawarden for trials, and moved to Hatfield for modifications on March 7th 1955, these being completed on November 30th 1956. It was issued to No.48 Maintenance Unit at Hawarden on December 4th and was to remain there until 1957. WX788 was declared a non-effective airframe on July 4th 1957 and was sold to Flint Technical College, at Connah's Quay, on November 25th 1957. Here it was used as an instructional airframe, and was to deteriorate somewhat. In 1972 it was purchased by the Chilterns Historic Aircraft Preservation Group and moved to their store at High Wycombe. In 1975 it was bought by SWAPS and moved to Rhoose. It is currently undergoing a full rebuild.

De Havilland DH.115 Vampire T. Mk.11 WZ425 '22'

Built 1952, constructor's number 15019, advanced trainer, powered by one De Havilland Goblin 35 turbojet of 3,500 pounds static thrust.

WZ425 was the twelfth production Vampire T. Mk.11 for the Royal Air Force, and was built at Christchurch. It was accepted from the makers off contract on August 28th 1952 and was delivered to the Central Gunnery School at Leconfield on September 4th. On April 14th 1953 it was transferred to No.43 Group's charge, this ppssibly being because the aircraft had been damaged. It was moved to No.27 Maintenance Unit at Shawbury on March 29th the following year. A year later, it was in service again, with No.229 Operational Conversion Unit at Chivenor. 1958, October 16th to be precise, found WZ425 back again at No.27 MU awaiting collection by another unit. A year later, it joined the RAF College at Cranwell, where it was given the nose-code '36'. From January 25th 1960 it was loaned to the Ministry of Aviation, before moving to RAF Oakington on June 29th, to join No.5 Flying Training School. Here it served with No.1 Squadron of the FTS, and was coded 'AN'. By about February 1961, when No.5 FTS changed is coding system, WZ425 became '22'. On September 26th 1963 WZ425 was retired to No.19 MU at St. Athan and was sold to Hawker Siddeley Aviation on November 25th 1967 and moved to Hawarden for storage. It later moved to Woodford. It was bought, via the British Aircraft Preservation Society, by SWAPS, and delivered to Rhoose on May 31st 1973 by road, it was the first aircraft at Rhoose.

Fairey Q Gannet AS. Mk.1

Built mid 1950s, three seat anti-submarine aircraft, powered by one Armstrong Siddeley Double Mamba coupled turboprop, not fitted.

This exhibit is the forward fuselage of a Gannet AS. Mk.1, its identity is unknown, and therefore so is most of its history. It is thought to be a spare unit, and that it has never seen operational service, however it is possible that it has flown at one time. It was bought from a Royal Navy store at Llangennech by a South Wales scrap merchant, and was probably in store from the days when No.849 Squadron was operating Gannets at Brawdy. It was bought from the scrappy by SWAPS and moved to Rhoose.

Hawker Sea Hawk FGA. Mk.6 WV826

Built 1954, single seat ground attack fighter, powered by one Rolls-Royce Nene turbojet.

This aircraft, and the Pembroke, below, are, at the time of going to press, the latest aircraft to join the collection. This Sea Hawk was in use with the School of Aircraft Handling at Culdrose, where it was used to teach aircraft handlers for aircraft carrier operations. It was given the instructional airframe number A2532 and the SAH indentity number SAH-2. It was delivered to SWAPS at Fairwood Common in later 1976 and at the time of writing is still there undergoing a respray prior to being installed in the Museum.

Percival Proctor V G-AHTE

Built 1946, constructor's number Ae.58, four seat tourer, powered by one De Havilland Gipsy Queen II in-line piston engine of 208 horse power.

G-AHTE was built at Luton in 1946, little is known of its flying career. It was withdrawn from use at Elmdon in August 1961 and slowly started to decay. Some time after this it was bought by a person in Llanelli, who moved it to his home, with the intention of restoring it to flying condition. It was bought by SWAPS in 1975 and has since been rebuilt and painted in typical War-time colours of an RAF communications Proctor.

Percival P.66 Pembroke C. Mk.1 WV753

Built 1957, communications aircraft, powered by two Alvis Leonides 127 radial piston engines of 540 horse power each.

This aircraft was delivered to Rhoose in early 1977 and is in near airworthy condition.

It is intended to keep the Pembroke in this state, so that it may fly again one day. WV753 ended its flying days in 1970 and was issued to No.4 School of Technical Training at St. Athan for use as an instructional airframe. It was allocated the maintenance airframe serial 8113M. It was disposed of by the SoTT in 1976 and bought by SWAPS.

Vickers 668 Varisty T. Mk.1 WJ944 'Y'

Built 1953, multi-purpose aircrew trainer, powered by two Bristol Hercules 264 radial piston engines of 1,950 horse power each.

This Varsity was last flown operationally with No.6 Flying Training School at Finningley, from where it was flown into the Museum on April 13th 1976. It is still in airworthy, but is not flown. WJ944 was built at Hurn and flew for the first time on February 19th 1953. It was delivered initially to No.8 Maintenance Unit at Little Rissington on March 11th and was to stay there, inactive, until May 4th 1956 when it was delivered to the Central Navigation and Control School at Shawbury, where it was coded 'A'. It moved on, with the same coding, to No.2 Air Navigation School at Thorney Island. It was than on charge with No.1 ANS with the code 'L' operating from Stradishall, followed by No.5 FTS at Oakington, coded 'M', back to No.1 ANS (this time coded 'Y') and then finally to No.6 FTS, still holding on to its code of 'Y'.

Vickers 732 Viscount 'G-WHIZ'

Built 1955, constructor's number 75, medium range airliner, powered by four Rolls-Royce Dart 506 turboprops of 1,547 shaft horse power each, not fitted.

The Wales Aircraft Museum is unique in having two Viscounts on display, albeit this example is only a fuselage. This display serves a dual purpose, being the Museum's Headquarters as well as an exhibit. G-WHIZ started life at Weybridge on June 7th 1955, when it first flew, as G-ANRS. Unlike G-AOJC below, G-ANRS has had a very varied and colourful past. It was delivered initially, on June 15th 1955, to Hunting Clan Air Transport. From October 7th that year it was leased to Middle East Airlines, Lebanon, as OD-ACH, returning to Hunting Clan (as G-ANRS) on September 30th 1957. In 1959 she was sold to the Egyptian airline, Misrair, and delivered to them at Cairo on August 31st as SU-AKY. In 1960 the airline was renamed United Arab Airlines, and SU-AKY was lettered accordingly. It was then bought by British Eagle International Airlines and delivered back to the United Kingdom (as G-ANRS) in March 1965. With British Eagle G-ANRS was named 'City of Newcastle'. From August 12th 1968 it was leased to Air France, but still carried the British registration, the lease terminating only three days later. Sadly British Eagle collapsed and G-ANRS became the property of Cambrian Airways in July 1969, the airframe being used only for spares for the remainder of Cambrian's fleet. The fuselage of G-ANRS was erected on a pedestal at Rhoose and used as a cabin trainer by Cambrian. In 1974/5 it was repainted in the colours of British Airways and some comical painter decided that G-ANRS was too mundane. and repainted the Viscount as 'G-WHIZ'. G-ANRS has a total of 23,405 flying hours.

Vickers 802 Viscount G-AOJC

Built 1956, constructor's number 152, medium range airliner, powered by four Rolls-Royce Dart 510 turboprops of 1,780 shaft horse power each, not fitted to the exhibit.

G-AOJC was originally allocated the registration G-AOHC, but this was not taken up. It first flew from Weybridge on November 11th 1956, and was delivered to British European Airways at Heathrow on January 19th the following year. It was given the name 'Robert O'Hara Burke' with BEA. On July 31st 1973 ownership was transferred to the new British Airways consortium, G-AOJC being allocated to the Scottish Division. After a very long, but uneventful, life serving the many routes of the giant corporation, G-AOJC was retired to the BA Viscount base at Rhoose, in 1975, ready for the scrap merchant. SWAPS rescued the aircraft, and it is now the major exhibit at the Museum. Restoration work is underway, and the Museum intend to repaint the aircraft in the colours of Cambrian Airways.

Westland WS-55 Whirlwind HAS. Mk.7 XG592

Built 1957, constructor's number WA/H/87, general purpose naval helicopter, powered by one Alvis Leonides Major 755 radial piston engine of 700 horse power.

First flown from Yeovil on April 24th 1957, XG592 was retained by Westlands for trials until being delivered to No.705 Squadron at Culdrose in November 1960, serving there until April 1962. It moved to Fleetlands and was stored until March 1963 when it was issued to No.846 Squadron, who used it in Malaya until 1964 when it was put into store at Sembawang. It returned to Fleetlands in 1967 and continued to be stored. It joined No.705 Squadron again in 1968 until retirement in December 1974 to the Royal Naval Aircraft Yard at

Wroughton. It was struck off charge in June 1976 and delivered to SWAPS at Rhoose that month.

Note : The following aircraft are also held by SWAPS and may well join the Museum in the future :-

Auster B5 AOP. Mk.9 BAPC.26 bare fuselage frame only, acquired from Berkswell. Possibly disposed of.
Slingsby T.7 Tutor BAPC.31 in store in Swansea.

LATE ADDITIONS

Such is the activity of the Preservation Movement in the United Kingdom that the following have occured during the typing of this manuscript.

Duxford

Predictably, the growing collection continues to grow, awaiting collection are a ***Dassault Mystere IV*** previously with the French Air Force from Sculthorpe, and the only other ***BAC TSR-2*** in existence (XR220, production number XO-4) has been procured from the Cranfield Institute of Technology, the last of a long line of historic aircraft held by the College. The other TSR-2 can be found in this book under 'Cosford Aerospace Museum'.

Strathallan

News from here is that another famous De Havilland type is to join the ranks, ***De Havilland DH.85 Leopard Moth G-AIYS*** in the near future.
There are also some planned exhibit changes as follows :-
Avro 652A XIX VM360 has been withdrawn from public display and is to be replaced by cousin *G-AHKX*.
Supermarine Spitfire Tr IX ML407 will be on display as just a fuselage, as will ***Short S.16 Scion II VH-UUP*** both will be untouched and will serve to show the amount of work put in by the staff at Strathallan to restore an aircraft to display condition.

APPENDIX

Despite considerable efforts, the compilers have been unable to extract sufficient information on the following 'museums'. In most cases the exact status of the museum is unknown, or details exactly as to the aircraft contents of the museum have failed to come to light. The compilers will of course welcome further details.

Kent Fighter Museum

Associated with the Battle of Britain Museum, Chilham Castle, Cantebury, Kent. Thought to have the following :-

Nord N.1011 Noralpha, identity unknown, in damaged condition, for disposal.
Vickers FB.5 Gunbus replica, identity unknown thought to have been built in 1976 by IES Projects Ltd of White Waltham for film use, possibly a flyer, in which case should be South African civil registered.

The Sea Fury at the Historic Aircraft Museum, Southend, WJ288, is due to move to the KFM.

Thorpe Water Park

Chertsey, Surrey. Here, in a leisure centre, run by Leisure Sport Limited, a small display of naval aircraft have been gathered. Several flying replicas have been built, or are being commissioned for displays both at Thorpe Water Park and at air shows etc. about the country. It is not known wether or not this section of the Park is open yet to the public. Aircraft for static exhibition are as follows :-

De Havilland DH.110 Sea Vixen FAW. Mk.2 XP919 arrived by road June 23rd 1975 from No.1 School of Technical Training at Halton, where it was 8163M, 'I-232'.
Hawker Sea Hawk FGA. Mk.4 WV798 '025' arrived by road in 1975 from the School of Aircraft Handling at Culdrose, where it was 'SAH-4'.
Westland WS-55 Whirlwind HAS. Mk.7 XM665 arrived by road from Lee-on-Solent December 16th 1974.

Aircraft available for display and flying are as follows (not based at Thorpe Water Park):-

De Havilland DH.82A Tiger Moth G-AOBX, floatplane.
Fokker Dr.1 replica G-BEFR, built by Rod Bellamy.
Supermarine S-5 replica G-BDFF/N220, built by the Aviation Design Bureau.

Thought to be underway also for the company are a Sopwith Camel replica and a Fairey Flycatcher replica (the latter was with Robinson Aircraft, now elsewhere, the former is with Rod Bellamy).

Warbirds Collection

Run by Doug Arnold at Blackbushe, Hampshire, a museum is intended to be opened here, and building is under construction. Actual date of opening is unknown, as are other details. Aircraft likely for eventual display are as follows :-

Boeing N2S-5 Stearman G-AWLO, under restoration, ex 5Y-KRR.
CASA 2.111 G-BDYA/T8B.124, ex Spanish Air Force, airworthy.
CASA 3.52 G-BECL 'N9-AA', ex Spanish Air Force, airworthy, in Luftwaffe markings.
Gloster Meteor TT. Mk.20 WM167, ex RAE Llanbedr, to be kept airworthy?
Hawker Sea Fury T. Mk.20 G-BCKH '252' in FAA colours, airworthy.
Hawker Sea Fury T. Mk.20 G-BCOV '253' in FAA colours, airworthy.
Hawker Sea Fury T. Mk.20 G-BCOW 'DW-A' in RAF colours (WW 2), airworthy.
SNCAN Stampe SV-4B G-AXPW, under restoration.
Sopwith Pup replica N5182/G-APUP owned by K. C. D. St. Cyrien, airworthy.
Supermarine Spitfire LF.XIV SL721/G-BAUP 'D-A', airworthy.
Tipsy Junior G-AMVP thought airworthy, camouflaged.

BIBLIOGRAPHY

The compilation of this monograph naturally took in many other works of reference, and the books listed below are the ones most refered to. Further reading on the subject of preserved aircraft is endless, there being a whole galaxy of possible sources, at many different levels. Below books used are listed by publisher, and are followed by journals and magazines used, again by publisher.

Books :-

Air-Britain (Historians) Limited

'Auster Aircraft', by N. H. Ellison & R. O. Macdemitria. Leicester 1966
'The DH.104 Dove & DH.114 Heron', by T. Sykes. Tonbridge 1973
'Impressments Log', Volume I, by P. W. Moss. London 1962
'Impressments Log', Volume II, by P. W. Moss. London 1963
'Impressments Log', Volume III, by P. W. Moss. London 1964
'Royal Air Force Serials K1000-K9999', by J. J. Halley. Billericay 1976
'Viking, Valetta, Varsity', by B. Martin. Tonbridge 1975

Avon-Anglia

'The Skyfame Aircraft Story'. Bristol 1976

British Aviation Research Group (including Blackbushe ARG)

'A History of the Armstrong Whitworth Meteor TT. Mk.20', Monograph No.2. Blackbushe 1969
'A History of the Blackburn Beverley', Monograph No.3. Blackbushe 1970
'A History of the Bristol/Westland Belvedere', Monograph No.2. Blackbushe 1969
'A History of the Douglas Skyraider AEW. Mk.1' Super-Monograph No.2. Hardmondsworth 1974
'A History of the Gloster Javelin Marks 7, 8 & 9', Monograph No.6. Blackbushe 1971
'A History of the Supermarine Scimitar', Monograph No.1. Blackbushe 1968
'A History of the Westland Wyvern', Super-Monograph No.1. Hardmondsworth 1973

Central Scotland Aviation Group

'Scotland Scanned 1975', by A. R. Benzies. Edinburgh 1975

G. T. Foulis

'Building Aircraft for "Those Magnificent Men in their Flying Machines"', by A. H. Wheeler. London 1965

Garnstone Press

'Veteran & Vintage Aircraft', by L. Hunt, fourth edition. London 1974

Ian Allan

'British Civil Aircraft Registrations', by G. Swanborough & J. W. R. Taylor. London 1970
'British Military Aircraft Serials 1911-1971' by B. Robertson. London 1971
'World's Airliner Registrations', by G. Swanborough, third edition. London 1971

A. K. Jenkinson

'Balloons around the World', by A. K. Jenkinson, first edition. London 1974
'Balloons around the World', by A. K. Jenkinson, second edition. London 1974

LAAS International

'Vickers Turbine Transports', by D. Dorman & A. Eastwood. Uxbridge 1974

Roger Lindsay

'De Havilland Venom', by R. Lindsay. Stockton-on-Tees 1973
'Service History of the Gloster Javelin Marks 1 to 6', by R. Lindsay. Stockton-on-Tees 1975

Merseyside Aviation Society Limited (including MAS, MSAE, MGAE)

'British Gliders', by P. H. Butler, second edition. Liverpool 1975
'British Homebuilt Aircraft since 1920', by K. Ellis. Liverpool 1975
'Irish Aircraft', by P. H. Butler. Liverpool 1972
'United States Air Force Serials 1946 to 1974', by P. A. Danby. Liverpool 1974
'United States Navy Serials 1941 to 1976', by P. A. Danby. Liverpool 1976
'Wrecks & Relics', by D. J. Stephens, first edition. Liverpool 1961
'Wrecks & Relics', by D. J. Stephens, second edition. Liverpool 1963
'Wrecks & Relics', by S. G. Jones, third edition. Liverpool 1968
'Wrecks & Relics', by K. Ellis, fourth edition. Liverpool 1974
'Wrecks & Relics', by K. Ellis, fifth edition. Liverpool 1976

Merseyside Aviation Society Limited/LAAS International
'British Independent Airlines since 1946', by A. C. Merton Jones, Volume 1. Liverpool 1976
'British Independent Airlines since 1946', by A. C. Merton Jones, Volume 2. Liverpool 1976
'British Independent Airlines since 1946', by A. C. Merton Jones, Volume 3. Liverpool 1976

Midland Counties Publications
'Beagle Aircraft - A Production History'. Hinckley 1974

National Air & Space Museum, Washington D.C., U.S.A.
'Aircraft in Museums around the World', Section I. Washington 1975
'Aircraft in Museums around the World', Section II. Washington 1975

Newark Air Museum
'Newark Air Museum'. Newark 1975

North East Vintage & Veteran Aircraft Association
'North East Aircraft Museum' Morpeth 1976

Pitkin Pictorials
'The Royal Air Force Museum - 100 Years of Aviation History', by Dr. J. Tanner. London 1975
'A Very Special Lancaster', by F. E. Dymond. London 1975
'Wings of the Eagle', by Dr. J. Tanner. London 1975

Putnam
'Aircraft of the Royal Air Force since 1918', by O. Thetford. London 1968
'Airspeed Aircraft since 1931', by H. A. Taylor. London 1970
'Armstrong Whitworth Aircraft since 1913', by O. Tapper. London 1973
'Blackburn Aircraft since 1909', by A. J. Jackson. London 1968
'Bristol Aircraft since 1910', by C. H. Barnes. London 1964
'The British Bomber since 1914', by P. Lewis. London 1967
'British Civil Aircraft since 1919', by A. J. Jackson, 2nd Edition, Volume 1. London 1973
'British Civil Aircraft since 1919', by A. J. Jackson, 2nd Edition, Volume 2. London 1973
'British Civil Aircraft since 1919', by A. J. Jackson, 2nd Edition, Volume 3. London 1974
'The British Fighter since 1912', by P. Lewis. London 1974
'British Naval Aircraft since 1912', by O. Thetford. London 1974
'Gloster Aircraft since 1917', by D. N. James. London 1971
'Miles Aircraft since 1925', by D. L. Brown. London 1970
'Shorts Aircraft since 1900', by C. H. Barnes. London 1967
'Vickers Aircraft since 1908', by C. F. Andrews. London 1969

Real Photographs
'The International Schneider Trophy Contests', by E. Hawks. Southport 1945

Royal Air Force Cosford
'Royal Air Force Cosford Aerospace Museum Guide Book', 1st Edition. Cosford 1973
'Royal Air Force Cosford Aerospace Museum Guide Book', 3rd Edition. Cosford 1975
'Royal Air Force Cosford Aerospace Museum Guide Book', 4th Edition. Cosford 1976

Royal Air Force Museum
'The Royal Air Force Museum 1973-1975'. Hendon 1975

Shuttleworth Collection
'The Shuttleworth Collection', 1964 Edition. Old Warden 1964
'The Shuttleworth Collection', 1975 Edition. Old Warden 1975

Strathallan Aircraft Collection
'Strathallan Aircraft Collection, Official Guide', 1st Edition. Strathallan 1974

Torbay Aircraft Museum
'Torbay Aircraft Museum, Souvenir Guide'. Torquay 1975

Journals and Magazines :-

Air-Britain (Historians) Limited : 'Air-Britain Digest' bi-monthly
'Le Trait d'Union' monthly
(journal of the French branch)

Air Yorkshire : 'Air Yorkshire' monthly

Anglia Aviation Society : 'Anglia Aeronews' monthly

British Aviation Research Group	: 'British Aviation Review'	monthly
	: 'Roundel'	monthly
Central Scotland Aviation Group	: 'Scottish Air News'	monthly
Michael Draper	: 'Auster Quarterly'	quarterly
Neville Franklin	: 'Control Column' (official journal of the B.A.P.C.)	six-weekly
Gatwick Aviation Society	: 'Hawkeye'	monthly
Alan W. Hall	: 'Aviation News'	fortnightly
Humberside Aviation Society	: 'Humberside Air News'	monthly
Ian Allan	: 'Aircraft Illustrated'	monthly
IPC Transport Press	: 'Aeroplane Monthly'	monthly
	'Flight International'	weekly
Merseyside Aviation Society Limited	: 'Flypast'	monthly
Midland Counties Aviation Society	: 'Air-Strip'	monthly
Popular Flying Association	: 'Popular Flying'	bi-monthly
Prestwick Airport Aviation Group	: 'Prestwick Air Letter'	monthly
Profile Publications	: 'Air Pictorial'	monthly
Gordon Riley	: 'Vintage Aircraft'	quarterly
South West Aviation Society	: 'South West Air News'	monthly
Ulster Aviation Society	: 'Ulster Air Mail'	monthly
West London Aviation Group	: 'South East Air Review'	monthly

THE B.A.P.C.

The British Aircraft Preservation Council was formed in 1967, at a meeting called by the Northern Aircraft Preservation Society (now The Aeroplane Collection) on the initiative of John Kenyon, their Chairman at that time. The Council meets quarterly at the home bases of member groups, and acts as a means of communication between aircraft preservation organisations, a channel of information to them, and a representative body able to speak on their behalf.

Close contacts established within the Council serve to reduce misunderstandings, and to prevent rivalries becoming jealousies. Whilst not seeking to own exhibits, the Council is frequently able to place exhibits in the hands of member organisations; exchange of material is also made much easier. As its status grows, the Council is able to apply pressure, or to speak on behalf of the preservation movement as a whole. Through affiliation to the Transport Trust, the Association of British Transport Museums and the International Association of Transport Museums, good relations are maintained with other organisations of like sympathies.

The Council has established the BAPC Register which is designed to assist historians and other interested parties by giving an identity to those aircraft which do not bear Service serials or qualify for inclusion in the Register of Civil Aircraft (many examples of aircraft registered with the BAPC can be found in the main text).

Membership is by election, and is available to any United Kingdom organisation or group with an active interest in aircraft preservation. Associate Membership is available to organisations based in the United Kingdom or elsewhere who wish to support the work of the Council.

Enquiries, offers of exhibits, or material of any kind, and donations for the general use of the movement, should be addressed to the Chairman, Peter Schofield B.Sc, 8, Greenfield Avenue, Urmston, Manchester, M31 1XN.

Member Organisations

Groups marked with an asterisk (*) are mentioned inside this monograph.

THE AEROPLANE COLLECTION : 100, Glazebrook Lane, Warrington, Cheshire, WA3 5BE.

AIRBORNE FORCES MUSEUM * : Browning Barracks, Aldershot, Hampshire, GU11 2DS.

AIRCRAFT PRESERVATION SOCIETY OF SCOTLAND : 28, Weavers Knowe Cres., Currie, Midlothian, EH14 5PL.

AIRCRAFT RADIO MUSEUM : J. F. Coggins, Baginton, Coventry, Warwickshire.

MUSEUM OF ARMY FLYING * : HQ, Army Air Corps Aviation Centre, Middle Wallop, Stockbridge, Hampshire, SO20 8DY.

BRITISH ROTORCRAFT MUSEUM * : Delta House, Summer Lane, Worle, Weston-super-Mare, Avon, BS22 OBE.

BATTLE OF BRITAIN MUSEUM : 2, Hillingdon Road, Gravesend, Kent.

CHILTERN HISTORICAL AIRCRAFT PRESERVATION GROUP : 29, Ash Close, Walters Ash, near High Wycombe, Buckinghamshire.

CORNWALL AIRCRAFT PARK (HELSTON) LIMITED * : Culdrose Manor, Helston, Cornwall, TR13 OGA.

DAN-AIR PRESERVATION GROUP : 50, Queen Mary Avenue, Basingstoke, Hampshire.

DUMFRIES & GALLOWAY AVIATION GROUP : 5, Terregles Street, Dumfries, DG2 9AA.

DURNEY AERONAUTICAL COLLECTION : 276, Weyhill Road, Andover, Hampshire.

DUXFORD AVIATION SOCIETY * : 116, Burstellars, St. Ives, Huntingdon, Cambridgeshire.

EAST ANGLIAN AVIATION SOCIETY : 15, Fishers Lane, Orwell, Royston, Hertfordshire, SG8 5QX.

FENLAND AIRCRAFT PRESERVATION SOCIETY : 'Rose Villa', Church Road, Christchurch, near Wisbech, Cambridgeshire.

501 BRISTOL AIRCRAFT PRESERVATION GROUP : 5, Stanley Crescent, Filton, Bristol, Avon.

FLEET AIR ARM MUSEUM * : R.N.A.S. Yeovilton, Ilchester, Somerset, BA22 8HT.

49 GROUP VETERAN AIRCRAFT ASSOCIATION : 26, The Ride, Kingswood, Bristol, BS15 4SY.

HISTORIC AIRCRAFT MUSEUM * : Aviation Way, Southend Airport, Southend, Essex.

HUMBERSIDE AIRCRAFT PRESERVATION SOCIETY * : 76, Pilgrim Ave., Immingham, S. Humberside.

IMPERIAL WAR MUSEUM * : Lambeth Road, London, SE1 6HZ.

LINCOLNSHIRE AVIATION MUSEUM * : Lancaster Farm, Tumby Woodside, Mareham-le-Fen, Boston, Lincolnshire.

MERSEYSIDE AVIATION SOCIETY LIMITED : 16, Walsingham Road, Childwall, Liverpool L16 3NT.

MIDLAND AIRCRAFT PRESERVATION SOCIETY * : 70, Broomfield Road, Earlsdon, Coventry.

R. J. MITCHELL HALL * : Kingsbridge Lane, Southampton, SO1 OGB.

MOSQUITO AIRCRAFT MUSEUM * : Salisbury Hall, London Colney, near St. Albans, Hertfordshire.

NEWARK AIR MUSEUM * : The Old Rectory, Irby, Laceby, Grimsby, Lincolnshire, DN37 7JR.

NORFOLK & SUFFOLK AVIATION MUSEUM * : 46, Lockhart Road, Ellingham, Bungay, Suffolk.

NORTH EAST VINTAGE & VETERAN AIRCRAFT ASSOCIATION * : 43, Glamis Avenue, Melton Park, Gosforth, Newcastle-upon-Tyne, NE3 55Y.

NORTHERN AEROPLANE WORKSHOPS : 'High Crimple', 1, Fulwith Grove, Harrogate, Yorkshire.

PERSONAL PLANE SERVICES LIMITED : Wycombe Air Park, Marlow, Buckinghamshire.

ROYAL AIR FORCE MUSEUM * : Aerodrome Road, Hendon, London, NW9 5LL.

ROYAL AIR FORCE COSFORD AEROSPACE MUSEUM * : Royal Air Force Cosford, Wolverhampton, West Midlands, WV7 3EX.

ROYAL AIR FORCE ST. ATHAN HISTORIC AIRCRAFT MUSEUM : Royal Air Force St. Athan, Barry, Glamorgan.

ROYAL SCOTTISH MUSEUM * : Chambers Street, Edinburgh, EH1 1JF.

SCIENCE MUSEUM * : Exhibition Road, London, SW7 2DD.

SHUTTLEWORTH COLLECTION * : Old Warden Aerodrome, Biggleswade, Bedfordshire, SG18 9ER.

SKYFAME AIRCRAFT MUSEUM * : Staverton Airport, Cheltenham, Gloucestershire.

SOLWAY AVIATION SOCIETY : 10, Percy Road, Carlisle, CA2 6ES.

SOUTH WALES AIRCRAFT PRESERVATION SOCIETY * : 3, Brynmead, Bryn, Llanelli, Dyfed, SA14 8QQ.

SOUTH YORKSHIRE AVIATION SOCIETY * : 15, Troon Road, Hatfield, Doncaster.

SOUTHERN AIRCRAFT RESEARCH ASSOCIATES : 5, St. Mary's Court, Eastrop Lane, Basingstoke, Hampshire.

STRATHALLAN AIRCRAFT COLLECTION * : Strathallan Airfield, Auchterarder, Perthsire, PH3 1LA.

SURREY & SUSSEX AVIATION SOCIETY : 59a, St. Helier Avenue, Morden, Surrey, SM4 6HY.

TORBAY AIRCRAFT MUSEUM * : Higher Blagdon, near Paignton, Devon, TQ3 3YG.

ULSTER FOLK & TRANSPORT MUSEUM * : Cultra Manor, Holywood, County Down, BT18 OEU.

VISCOUNT PRESERVATION TRUST : 17, Portsmouth Road, Thames Ditton, Surrey.

YORKSHIRE AIRCRAFT PRESERVATION SOCIETY : 4, Thorsfield Ave., Muncaster, York, YO3 9LP.

Associate Members

'Aeroplane Monthly', Dorset House, Stamford Street, London SE1 9LU.
The Moth Club of the Antique Airplane Association (U.S.A.), Kalamazoo, Michigan, U.S.A.
'Aviation News', 2, Sheepfold Lane, Amersham, Buckinghamshire.
The Brooklands Society Limited, 21, Heathpark Drive, Windlesham, Surrey.
Derby Industrial Museum, The Silk Mill, off Full Street, Derby, DE1 3AR.
*Irish Aviation Museum *, 'Spindrift', Shore Road, Portmarnock, County Dublin.*
Loughborough Leicester Air Museum Preservation Society, 50, Tuckers Road, Loughborough.
Model & Allied Publications Limited, PO Box 35, Bridge Street, Hemel Hempstead, Herts.
Second World War Aircraft Preservation Society, 4, Burns Close, Basingstoke, Hants.
Society of Licensed Aircraft Engineers & Technologists, 'Grey Tiles', Kingston Hill, Kingston-on-Tames, Surrey.

THE M.A.S.

The Society was formed in 1956 to stimulate interest in aviation on Merseyside and in the country in general. The M.A.S. is one of the oldest, yet one of the most progressive, of the aviation enthusiast organisations in the U.K. In 1975 it became a Company Limited by Guarantee and can presently boast just under 700 members.

Some facets of the M.A.S. :-

'Flypast' The M.A.S. is unique in publishing *and* printing its own litho magazine - 'Flypast'. This flourishing magazine now average 22 sides per issue, and includes at least one page of photographs and regular line-drawings. Emphasis is on a detailed review of all aspects of aviation in the north west but coverage on a national and international scale can also be found within its pages. 'Flypast' is fast becoming a 'must' for many enthusiasts.

Clubroom Situated at Liverpool Airport, open each weekend, the venue of film and slide shows held on a regular basis. Books and refreshments available.

Trips The M.A.S. runs a comprehensive selection of trips to suit all tastes. Coach trips to air shows up and down the country; charters to overseas air displays; air experience flights in various types of aircraft. Details appear in the Society's journal 'Flypast'.

Aircraft Preservation The Society's first venture in this field was the preservation of Avro Avian G-EBZM which ultimately led to the formation of the Northern Aircraft Preservation Society (now The Aeroplane Collection). Current work involves the restoration of Fa 330 gyro-kite W/No. 100549 and the long-term storage of Vampire T. Mk.11 WZ553. The M.A.S. is an active member of the British Aircraft Preservation Council and of the Transport Trust.

Membership Is open to anyone interested in aviation. For current details of membership and a sample copy of 'Flypast', send a stamped addressed A4/foolscap size envelope to the Registrar, at the following address :-

Mr. P. Duckett, Registrar, M.A.S., 265, Rose Lane, Liverpool L18 5HJ, Merseyside.

Publications

The Society has established itself as a major publisher of enthusiast publications on a wide range of subjects. Amongst those currently available are :-

Wrecks and Relics
Now in its 5th edition, this is our very popular guide to preserved, instructional, derelict and dumped aircraft in the United Kingdom and Eire. ***This is an ideal companion to 'British Museum Aircraft'.*** It delves into every nook and cranny of the country to reveal in considerable detail the fates and fortunes of non-flying aircraft. A fascinating study, over 500 locations mentioned, 60 well-chosen photographs, two indices and seven appendices, all bound into 112 pages of invaluable reference work. Price £1.70 including postage.

British Independent Airlines since 1946
Four massive reference works on the airlines that have forged routes and services since 1946. Hours of fascinating reading, hundreds of superb photographs.

Volume One,	Ace Freighters to British Westpoint,	price £2.40 including post.
Volume Two,	Brooklands Aviation to Irefly,	price £2.40 including post.
Volume Three,	Irish Air Charter to Sagittair,	price £2.50 including post.
Volume Four,	Scillonia Airways to Yorkshire Aeroplane Club ,	price £2.50 including post.

British Isles Airfield Guide
Now in its 7th edition, our popular guide to the airfields of the United Kingdom and Eire, operational and disused. A handy companion. Price 65p including postage.

British Airfields Past and Present
A superb three-colour wall chart of Britain and Eire's aifields. Produced on high quality glossy paper, will grace any wall. Full index on reverse. Sent by post, with two folds, 27p for one, 37p for two, including postage.

Orders for the above, all monies payable to Merseyside Aviation Society Ltd.
There are several other titles available, and others always being launched, watch the aviation press for details. Or send an SAE to the address below for the latest list.

M.A.S. Publications, Dept. K, Room 14, Hangar 2, Liverpool Airport, Liverpool L24 8QE.

AIRCRAFT INDEX

By aircraft type, alphabetically, some abbreviation of type names has been necessary. Page numbers in italics, e.g. *93*, refer to an aircraft mentioned only in brief in the 'Notes' at the end of an entry, i.e. incomplete airframes, aircraft not on public view, or in store.

space for notes.....

space for notes.....